BEYOND THE MAP

BEYOND THE MAP

Unruly Enclaves, Ghostly Places, Emerging Lands and Our Search for New Utopias

ALASTAIR BONNETT

THE UNIVERSITY OF CHICAGO PRESS

CHICAGO

The University of Chicago Press, Chicago 60637
© 2018 by Alastair Bonnett
All rights reserved. No part of this book may be used or reproduced in any manner
whatsoever without written permission, except in the case of brief quotations in critical
articles and reviews. For more information, contact the University of Chicago Press,
1427 E. 60th St., Chicago, IL 60637.
Published 2018
Printed in the United States of America

27 26 25 24 23 22 21 20 19 18   1 2 3 4 5

ISBN-13: 978-0-226-51384-3 (cloth)
ISBN-13: 978-0-226-51398-0 (e-book)
DOI: https://doi.org/10.7208/chicago/9780226513980.001.0001

First published in the United Kingdom in 2017 by Aurum Press, an
imprint of The Quarto Group

Quotes on pages 195–96 from *Marshland: Dreams and Nightmares on the Edge
of London* reproduced with permission from Influx Press. © Gareth Rees, 2012,
Marshland: Dreams and Nightmares on the Edge of London.
Quote on page 194 from *Scarp* by Nick Papadimitriou reproduced
with permission from Hodder and Stoughton.
Illustrations © Rachel Holland

LIBRARY OF CONGRESS CATALOGING-IN-PUBLICATION DATA
Names: Bonnett, Alastair, 1964– author.
Title: Beyond the map : unruly enclaves, ghostly places, emerging lands
and our search for new utopias / Alastair Bonnett.
Description: Chicago : The University of Chicago Press, 2018.
| Includes bibliographical references and index.
Identifiers: LCCN 2017044379 | ISBN 9780226513843 (cloth : alk. paper)
| ISBN 9780226513980 (e-book)
Subjects: LCSH: Geography—Miscellanea. | Curiosities and wonders.
Classification: LCC G123 .B66 2018 | DDC 910—dc23
LC record available at https://lccn.loc.gov/2017044379

♾ This paper meets the requirements of ANSI/NISO Z39.48-1992
(Permanence of Paper).

Contents

Introduction

Geography is getting stranger: new islands are rising up, familiar territories are splintering and secret realms are cracking open their doors. The world's unruly zones are multiplying and changing fast.

I present thirty-nine stories from thirty-nine extraordinary places, each with something to tell us about the shifting nature of place and place-making. We'll be meeting warring enclaves and contemporary utopias, along with many other geographical outliers and pariahs. They are all unique but also connected: the uncanny ruins, the unnatural places, the escape zones and gap spaces, are sites of surprise but also of bewilderment and unease. The compass needle is starting to spin. The adventure before you responds to a new spirit of geographical giddiness. The dizzying fragmentation, the overlapping and shape-shifting of borders that is seen in so many parts of the planet, tell us that geography is not a staid and dusty affair but is enthralling and often alarming.

1

I've chosen places whose unique and intriguing stories deserve to be better known and that tell us something about our present state of geographical turbulence. Some, like Trap Streets and the Ferghana Valley, came about as a result of suggestions from readers of my previous books, others from my own out-of-the way travels and research.

We start with unruly islands: disputed, fantastic and legion, they are cresting waves around the world, from the chilly English Channel to the warm seas of the Philippines. Next we explore some old allies of the island: enclaves and new countries. In deep valleys in the Italian Dolomites, and behind a wall that stretches across the Sahara, can be found very different types of survival struggle. These strange and uncertain new territories are brimful with bold ambitions. It's a small leap to our next destination: utopian places. The fraying of traditional geographical borders and loyalties is unleashing utopian energies of every stripe, from the darkest to the most playful. As well as ripping up the map, the radical Islamists of Syria and Iraq are creating a version of utopia, a genocidal exemplar of utopia's desire for pure territory. Thankfully there are many other ways of seeking perfection: from beautiful Christiania with its acres of higgledy-piggledy self-build, to the movable pods and capsules of the 'new nomads', an affluent, self-contained tribe of permanent travellers. Dystopia may get all the cultural hype but in the real world, even around the corner, people are busy trying to build workable alternatives.

I've been trying to work out why I'm so obsessed with place. It's certainly not just because I'm a professor of geography. If I'm honest, it's not a cerebral thing at all. It's to do with the glee and the drama, the love and the loathing – the powerful emotions that are poured into place. That's not enough either: for I am afflicted with an aching nostalgia for lost places, a yearning for the disappeared that guides my thirty-nine steps. In the second half of the book I take you to ghostly and hidden places; from an abandoned British graveyard in India to Garbage City in Cairo; from the hidden zones off Google Street View to the mysteries of the Tokyo metro system; all places that are as unique as they are profoundly disorientating.

I guess I'm so drawn to the concealed and the spectral because, like so many of us, I'm surrounded by a sense of the fleeting and the impermanent. For as long as I can remember, much of the landscape has looked and felt like a building site. On the long roads I drive, the pathways to work, on each side I'm surrounded by streets being gouged up, shovelled away and ever more elaborate traffic systems and flimsy, shed-like buildings being bolted into view. In response, shy hints of the past, leftovers and remnants, take on a totemic power.

Why is that? Ordinary reasons, yet they are mine. Visiting my granny as a small boy, in her chilly house surrounded by overgrown fields, I'd wrinkle my nose at the thick scent of coal tar soap, mothballs and door blankets; hear the ticking clock above the snapping coals in the fire. I was swaddled in the past. We'd drive to her remote Suffolk village across sandy heaths, and pass row upon row of neat new terraced homes behind the high wire fences of the adjacent US Air Force base. The serried lines of back garden barbecues, as well as the base's substantial pile of 'free-fall nuclear bombs', all seemed primed and ready to go. The contrast between the fuggy comfort of granny's home and the prospect of Armageddon, of the winding down, the falling away of village life and the muscular spectacle of world-shaping power, lodged itself in my heart and my geographical imagination.

That was getting on for fifty years ago, but I guess it still helps to explain why I'm so drawn to places that have an uneasy mixture of the past and present. Many of the places I explore in this book exhibit a similar layering. Modern places might sometimes feel like leaps into gloriously empty air, but the ghosts won't go away, and I've come to accept them as necessary, even hopeful, presences. Clambering through the ruins of the Boys Village, which lies in the shadow of a power station in South Wales, a seaside campus for boys from mining communities now collapsing with the weight of weeds and covered in graffiti, I found myself searching out the fireplaces, half-expecting the cosy smell of mothballs, the signs of comfort in an unloved landscape.

A place is a storied landscape, some*where* that has human meaning. But another thing we have started to learn, or relearn, is that places aren't just about people; that they reflect our attempt to grasp and make sense of the non-human; the land and its many inhabitants that are forever around and beyond us. It can be an unnerving exchange, especially when what we hope to see is something purely natural, and what we find instead is our own reflection. Shorelines are waxing and waning with increasing speed, and old kingdoms, like Doggerland, as well as new ones in the once-inaccessible Arctic, are being revealed, demanding that we look at the land-scape, and at the map, in new ways; as something in motion, unmoored by tradition.

The ever more unruly maps of human and physical geography can seem overwhelming. Perhaps that's why little places – the small secrets, the hidden surprises – have come to feel so important. Some of these retreats are simple delights, like Nek Chand's Rock Garden, a labyrinth of lanes, waterfalls and tens of thousands of sculptures of wide-eyed humans and prancing animals, all fabricated from the detritus of the traffic-clogged new town that surrounds it. Others are rather less happy, such as the anti-pedestrian interzones between ever-churning roads. But they all share a sense of unruly autonomy, of having escaped from the iron cage of the ordinary. These fragments are, in their own odd way, stabs at utopia. This book starts with 'Unruly Islands', and I end that first part with the same lonely traffic island, a scrap of unnamed land wedged between highways, that I devoted a chapter to in *Unruly Places*. This time I'm carrying across a bag of compost and a few fruiting plants. It's my fragment of Eden, my silly gesture against the din of traffic. I like to think that one day soon I might catch a fleeting glimpse of strawberries, pushing through the thin soil in spite of the chaos that surrounds them.

1.

Unruly Islands

Here are six of the world's most surprising islands and six surprising stories. Each island (or group of islands) unsettles the complacency of the mainland; not least what could be considered Britain's farthest southern shore, the islands of Les Minquiers off Jersey, and the US's uncertain, fabulously remote and disparate archipelago, the Minor Outlying Islands. The line between fact and fiction can be very thin in small islands; nowhere more so than in the ones that are being rebuilt. The South China Sea's Spratly Islands have been transmuted over recent years from a scatter of pristine reefs into weaponised and squared-off fortresses. Human hubris is often writ large on the most fragile islands, which is why it is worth being reminded that the most fundamental planetary forces are entirely beyond our control. The post-glacial 'bounceback' of land seen across the high north, such as Bothnia's rising islands, is sprouting umpteen new shores from the sea, whether we like it or not. It is not clear that we can even count them all. The problem of counting islands is one I turn to in the company of 534 'new islands' recently discovered hiding off the coast of the Philippines. Islands turn out to be more elusive, cartographically speaking, than one might suppose; none more so than the ones cradled in the indifferent arms of grinding roads, such as the traffic island I visit at the end of this part of the book, taking with me some wild strawberry plants.

Les Minquiers

I'm waiting on a gently bobbing pontoon with a gaggle of other brightly life-jacketed holidaymakers for a motor boat to take us to the most southerly part of the British Isles, a place whose sovereignty was only finally secured in 2004. It's a cloudless day in early April, and soon we're gripping on tight as the inflatable roars us far from any shore, skipping across the bright water about 14 miles south of Jersey's capital, Saint Helier. After twenty-five minutes, a galaxy of sharp crags pierces the horizon. Les Minquiers, which people on Jersey call 'the Minkies', stretch across an area considerably larger than Jersey itself, and at low tide expose 77 square miles of sand and rock (Jersey is 46 square miles).

It's the uncertain scale of Les Minquiers that intrigues me. A couple of times a day this is a vast place, but at high tide only nine islets are visible, and only one of these is of any size at all; called La Maîtresse Île, at its smallest it is a mere 328 feet long and 164 feet wide. The tidal range in this part of the

English Channel is huge, up to 40 feet. The Minkies appear and disappear before your eyes, something pulled from nothing; a magic archipelago.

The suntanned and affable skipper pulls down on the throttle and a watery silence engulfs us. We've swerved round before La Maîtresse Île. A row of small one-storey stone houses is crammed along its single ridge, jostling together to keep their toes out of the waves. Edging carefully along the slippery, seaweed-strewn pier, my first port of call is the island's outside toilet. It stands boldly out on its own, and is the most southerly building in Britain, a plaque on its flapping door proudly proclaiming this unique distinction. To make use of the loo requires hauling up a bucket of seawater to sluice the pan and, in any case, I don't have time to tarry; the tide is turning, and soon the boat won't be able to make it to the pier.

The empty stone huts are drizzled with white bird poo. There are twelve of them, ten of which are owned by Jersey families and occasionally used for weekend lets; the other two are owned by the States of Jersey. One of these is the customs house, marked by an incongruously grand carved stone bearing the three lions of Jersey and the words 'États de Jersey' and 'Empôts'. Over at the far side of the island, there's a weathered helipad. But my eyes keep getting drawn down to the ground. The rocks and ground-hugging, large-leaved scrub are covered in black and red firebugs. They scuttle about frantically, as if searching for something lost.

Once back in the boat, our skipper tells us that this is 'the largest unmapped area in the Western world'; its great tidal variation meaning that only local knowledge can find a way through the islets beyond La Maîtresse Île, a zone called the Wilderness. I'm feeling lucky; the weather here can be foul, but right now the sea is full of colours: translucent azure and soft greens, graceful shades that shoal around the islets and white sandbanks. On a warm, sunny morning it's a tempting place, and I can understand why Jersey people will motor or even paddle out to their own favourite spots, where they can have an island all to themselves.

The landscape is draining so quickly that what looked, at first glance, to be a scene of isolated rocks poking from the sea is transforming into a place of lagoons and little rocky hills connected by sweeping dunes. The bottom of our inflatable softly nudges its way onto a sandbar and we hop out, squelching into the virgin sand. Narrowing my eyes against the sun's glare on this glittering and ephemeral island, it seems odd to think that somewhere so lost could have such a long and contested history.

In 1792, the reef began to be quarried for granite, which was ferried back to Saint Helier. The main island's stone huts date from this period. Jersey fishermen, who prized the area's rich catch and used the island as a base, apparently brought the quarrying to a stop by throwing the workmen's tools into the water. However, it is the fact that Les Minquiers exist on an uncertain boundary between France and Britain that has been the most persistent source of conflict. Being almost as near to France as they are to Jersey, which is itself far closer to France than to England, it is not surprising that the French have long argued that this reef is theirs. In April 1938, the French prime minister Édouard Daladier thought the issue so important that he landed on La Maîtresse Île to assert the French claim.

Bigger geopolitical issues soon pushed the mastery of the Minkies down the agenda. During the Second World War the Germans had an observation post on La Maîtresse Île. Being stationed out here must have felt like leaving the planet. The few German soldiers marooned on this windswept spot ended up being forgotten and bypassed by the war. In *The End of the War, Europe: April 15–May 23, 1945*, historian Charles Whiting reports that on 23 May 1945, over two weeks after the war in Europe had ended, Lucian Marie, captain of the fishing boat *Les Trois Frères* 'on watch on the bridge, suddenly became aware that the island – a collection of low reefs – was inhabited'. An armed German soldier emerged. 'Listen, Frenchman,' the German said, 'we've been forgotten by the British. Perhaps no one in Jersey told them we were here. So, now we've had enough. We are running out of food and water. You must help us.' 'How?' asked Lucian Marie. 'Simple, I want you to take us over to England,' came the reply. 'We want to surrender.'

The war was finally over for these forgotten soldiers, and they no doubt fondly hoped never to see the Minkies again. However, the French claim on this lonely archipelago was soon to bob up again. In 1953 the case of both Les Minquiers, and a similar range of islets off the north coast of Jersey called Les Écréhous, was sent to the newly established International Court of Justice for arbitration. France's case rested on its proximity and its tradition of fishing both areas. Britain's claim focused on its building and occupation of the stone huts. The latter viewpoint swayed the judges, and it was announced that the sovereignty of both reefs 'belongs to the United Kingdom'.

Since one of the first things that any visitor to Jersey learns is that the island is not in the UK, and the International Court of Justice did not determine the boundaries of the outlying reefs, the 1953 judgement did not settle the dispute. Certainly it didn't convince some critics in France. These included the novelist Jean Raspail, an eccentric but determined nationalist whose best-known work is *The Camp of the Saints*, a novel that predicts an invasion of migrants from 'the South' engulfing and destroying Western civilisation. In 1984 Raspail sailed to Les Minquiers and hoisted a Patagonian flag, an ironic gesture against Britain's ongoing attempt to recapture the Falkland Islands from Argentina. Twelve years later Raspail returned to La Maîtresse Île and took down the British flag, which he then presented to the British ambassador in Paris. To the north, the Les Minquiers' sister islands were seeing similar acts of symbolic appropriation. In 1993 and 1994, Norman flags were raised on Les Écréhous by French 'invaders'.

It was eventually decided that the 1953 International Court of Justice decision needed to be revisited, not because of all the flag-waving, but because these reefs' queasy topography, with its huge daily variation between land above and below the water, meant that a far more detailed delimitation was needed. It took thirteen years of talks between France and Britain to arrive at an agreement. One Jersey politician involved in the negotiations described it as literally counting the Minquiers and Écréhous 'rock by rock'. By 2000 new political maps could be issued showing what it was now hoped was the definitive maritime boundary line between Britain and France. This agreement, along with another document detailing fishing grounds, came into force on 1 January 2004. Soon after, buoys were deployed to physically mark out the various lines in the water that, finally, separated Britain from France.

All of these geopolitical manoeuvrings seem a million miles away from the still and silent sandbar that I find myself on. With every passing minute it grows a new shoreline. The silky wet sand is rivuleted with braided streams that tumble out from the island's humped golden spine. I'm lulled; lying down now; feeling sleepy in the hot sun. All the world's water is washing down a plughole and soon it will all be gone. But that fantasy provokes its alarming opposite; for the tide will soon turn, and I must wake up and sail away to somewhere safe, somewhere certain. I crane my neck for reassurance: there at the end of this unnamed, unnameable island is the boat to take me home. I know already that I'll carry the memory of Les Minquiers as something between a dream and a feeling of foreboding.

The United States Minor Outlying Islands and the United Micronations Multi-Oceanic Archipelago

This story starts in obscure places and ends up in an outlandish one. The United States Minor Outlying Islands are the least-known bits of the US. In total these tiny and utterly remote islands cover just 13 square miles. There are nine of them, eight in the Pacific (Baker Island, Howland Island, Jarvis Island, Johnston Atoll, Kingman Reef, Midway Atoll, Palmyra Atoll and Wake Island) and one in the Caribbean Sea (Navassa Island). They are a collection of oddments. Their collective title 'United States Minor Outlying Islands' is a label of convenience, since the islands have no government. They are administered as National Wildlife Refuges by the US Fish and Wildlife Service, except for Wake Island, which is run by the US Air Force.

All but Wake Island were claimed under the Guano Islands Act of 1856. This high-handed piece of American legislation proclaims that

> Whenever any citizen of the United States discovers a deposit of guano on any island, rock, or key, not within the lawful jurisdiction of any other Government, and not occupied by the citizens of any other Government, and takes peaceable possession thereof, and occupies the same, such island, rock, or key may, at the discretion of the President, be considered as appertaining to the United States.

'Guano' is a Quechua word for agricultural dung. Seabird guano, which is what the island hunters unleashed by the Guano Islands Act were after, contains high levels of nitrogen, phosphate and potassium, and is the world's most prized natural fertiliser. The Act, which is still in force, led to claims on about one hundred islands dotted across the world. Most of these claims were not actively defended after the guano was dug out, and were either rescinded or became dormant. An example of a rescinded claim are the Swan Islands, a group of three islands off the east coast of Central America, which were ceded to Honduras in 1972. A case of a dormant claim is Ducie Island, which is in fact a group of uninhabited islets that cover

170 acres and lie 332 miles east of the Pitcairn Islands, the only remaining British Overseas Territory in the Pacific. Since 2010 Ducie has been formerly claimed by the British as part of the Pitcairn Islands.

France and the UK have collections of small, far-flung islands. But the United States Minor Outlying Islands is singular in its legal oddity. Christina Duffy Burnett, a professor of law at Columbia University, has long been intrigued by the indeterminate status of these scattered flecks of America. They are, she says, 'a weird sort of non-place, from a constitutional perspective'. In her opinion the 'islands "belong" to the US, but they aren't really a "part" of the United States', and so she asks, 'What law applies there? Not really so clear.'

It would be a mistake to imagine that because they are distant and tiny they are unimportant. Each of the islands allows the US to claim vast tracts of ocean as part of the 200-nautical-mile 'exclusive economic zone' that extends from any country's shore. Moreover, each island has its own story. Two of the names in the list of Minor Outlying Islands will leap out to students of twentieth-century military history: Johnston Atoll and Wake Island.

Wake Island is the only one of the group that has a resident population, comprising ninety-four US military personnel. A U-shaped coral atoll, the island was invaded by the Japanese on the same day that they attacked Pearl Harbor. Wake Island fell on 23 December 1941 after fierce battles that took nearly a thousand lives. Once back under US control, the island resumed its military function and is today used for missile tests and as a refuelling stop.

Johnston Atoll is made up of four flat, sandy islands, the largest being Johnston Island. Johnston Island has been artificially bulked out, growing from 46 to 596 acres, in part to accommodate a longer landing strip. Today it is a long, unnatural-looking rectangle. At its peak about a thousand personnel were stationed there. The atoll was used for nuclear weapons testing in 1962 and for rocket launches, and has a 25-acre landfill full of toxic material, including drums of Agent Orange from the Vietnam War. To add to the poisonous brew, in the 1990s the island hosted an incineration plant for chemical weapons, among them sarin nerve gas.

The last soldier left Johnston Atoll on 17 August 2001. Given that it is little more than a dumping ground for toxic waste it was a surprise that, in July 2006, Johnston Island was listed for sale by the US Government's General Services Administration (GSA) as a 'residence or vacation getaway' with potential usage for 'ecotourism'. Perhaps the GSA enjoys irony. The listing was later described as a 'teaser' being used to gauge commercial interest.

There is another and lesser-known story about these islands. The school records of Kamehameha Schools in Hawaii describe surprisingly dedicated attempts by its pupils to colonise Howland, Jarvis and Baker Islands. The colonisation process began in 1935, and the accounts have a fresh and optimistic flavour, with photos and local newspaper headlines about the smiling young adventurers. Sadly the adventure ended in tragedy. The December 1941 diary entry from the boys on Howland Island takes up the story.

Suddenly Joe Keliihananui looked up and saw 14 twin-engined bombers flying in high from the north west . . .

From a height of about 10,000 ft. the bombers let us have it. They dropped about 20 bombs, then turned and came back over the islands, dropping some ten more. The explosions shook the ground under our feet and the smoke concealed almost everything from our view.

When the planes finally left, Mattson and I walked over to where Dick and Joe were lying. They had been badly hit. They were both hurt in the legs and one had a chest wound and a hole in his back. We were going to fix up a place to put them, but by the time we got something arranged, they were dead.

By 1 January 1942, nearly a month after the first attack, the boys 'were convinced that we were in the middle of this war's no man's land and that we would probably have to stay there for the duration'. Thankfully they were rescued by an American destroyer on 31 January.

The Guano Islands Act is a legal oddity that people stumble across and start fantasising about. There are plenty of online stories that run under headlines such as 'Thanks to a 19th century law, Americans can lay claim to any uninhabited island with birdshit on it'. They are invariably followed by chat forums in which an initial excited optimism is deflated, contributors concluding that there aren't any unnoticed, unclaimed and unoccupied islands out there waiting for new residents. I'm not so sure about that, but I'm just as interested in the question of why any one country's claims should be respected. The US's

Guano Islands Act is not international law, and its legal status is questionable.

If the US can claim numerous empty islands for itself, then group them together into a loose federation of 'Outlying Islands', then what is stopping anyone else from doing the same? This question helps us get from the Minor Outlying Islands to the outlandish concoction that claims to have annexed them and many other remote spots, the United Micronations Multi-Oceanic Archipelago (UMMOA). To give a flavour of this entity it is helpful to note that it was founded on 19 January 2008 by the Most Rev. Dr Cesidio Tallini, an 'alternative scholar' and serial creator of micronational entities. The UMMOA leaves a dense trail in the virtual world of microstates, in part because of the scale of its ambition. It doesn't stop at claiming all of the Minor Outlying Islands but has twenty-nine territories, mostly scattered reefs and islets that have 'no indigenous population', as well as a chunk of the Antarctic.

The UMMOA, which claims to have sixty-eight 'nationals', is buoyed by rafts of provocative statements about its global ambitions. These include extending its territorial claims to unwanted and despoiled realms, such as 'a piece of the Great Pacific Garbage Patch' as well as space debris and several islands disappearing as a consequence of rising sea levels. The UMMOA has also tried, unsuccessfully, to host a micronational version of the Olympic Games. Dr Tallini reports that 'Unfortunately, the other people did not prove to be active or reliable.'

The UMMOA assures us that it is not 'an ego trip, and is actually carried on through the multifaceted work of adults'. Despite such curious claims, I'm attracted by the UMMOA's reinvention of the idea of the 'archipelago'. An archipelago is a collection of islands where the dots are joined; its constituent islands have a relationship across water; lines of connection weave them together. Dr Tallini's 'Multi-Oceanic Archipelago' is no more real than his doctorate in 'cyberanthropology', but his invention highlights one of the reasons why the Minor Outlying Islands, as well as his own, only slightly more unlikely, creation are so alluring. Despite their rather depressing history of being mined-out and poisoned, these tiny, unvisited places are thousands of miles apart, yet they have been chained into a kind of fabulous community. They form a weightless concoction; something unfeasible and spun loose from the ordinary world.

The New Spratly
Islands

Building an island isn't so very difficult. If you choose
your spot well you'll get an island if you just keep shov-
elling on. By far the world's largest artificial island is the 375
square miles of Flevopolder in the Netherlands, which was
built in the 1950s and 1960s. It's now covered in smooth,
peaceful fields and villages. Elsewhere political, economic
and environmental pressures are creating altogether more
spiky examples. The warm waters of the South China Sea
are framed by populous nations, each with competing claims
over the hundreds of islets and reefs that make up the, until
very recently, pristine and uninhabited Spratly Islands. Today
the Spratly Islands are being bulked out, squared off, covered
in concrete and turned into offensive military bases. This
once-unspoiled tropical paradise has been transmuted into
an army of geographical Frankensteins.

The Spratly Islands comprise thirteen islets with a land area above two acres, but mostly it is a sprinkle of reefs and crags. In their natural state, the shifting patterns of sand on the larger islands are sculpted and shifted anew each season by wind and intense monsoon rains. Spread over 164,000 square miles (425,000 square kilometres), the Spratly Islands are made of coral and support little plant life and have no fresh water. But the warm shallow seas around them are home to a huge variety of marine creatures, including rare turtles and thousands of species of fish.

It was nine o'clock in the morning on 29 March 1843 when London-born captain Richard Spratly sighted the low, sandy shore of a small island that he unimaginatively named 'Spratly's Sandy Island'. There not being anything else to report, he promptly sailed off. Though perhaps the captain's later travails were a presentiment of what was to come. The following year he was accused of duplicity and intrigue and of abandoning some of his men on an island in the Sea of Japan. A few decades later, another British captain claimed the Spratly Islands for himself, calling it the Republic of Morac-Songhrati-Meads. His descendants divided into two competing clans, one renaming the Spratly Islands the 'Kingdom of Humanity'. The reunited kingdom was petitioning neighbouring states and even trying to sue the US for a deed of ownership even up to the 1980s. Yet another pretender to the Spratly throne was a Filipino called Tomás Cloma, who sailed to the Spratly Islands in 1956 and designated them 'Freedomland'. This innocuous gesture provoked Taiwan to send a military force to occupy the largest island in the chain, the 4,590-feet-long Itu Aba Island.

In 1978 Cloma sold his entire kingdom to the Philippines for one peso, from which point on the Spratly Islands were claimed as Philippine national territory.

The playful stories of these eccentric claimants sit oddly with the far more sombre contemporary history of the Spratly Islands. Because of its snarling, jagged reefs, the southern area of the South China Sea is called Dangerous Ground, but it's a name that could be applied to the whole region. More than US$5.3 trillion worth of shipping travels through the South China Sea each year. These waters also contain extensive and untapped oil and gas reserves and about 12 per cent of the world's fish catch. Sitting in the middle of all of this, the Spratly Islands are at the crossroads of multiple high-stakes ambitions, and each little reef and islet is today covered with layers of territorial claims and counterclaims.

The five local protagonists are China, the Philippines, Malaysia, Vietnam and Taiwan. The US is also a key player: China's ambitions to weaponise the Spratly Islands have not gone unnoticed in Washington. There has been limited land reclamation in the past, the most sizeable being on Swallow Reef, which was transformed from 15 acres to over 86 acres by the Malaysians. The Philippines is the only claimant that has not yet indulged in island-building. Although there have long been plans to extend and remodel the islands that it has de facto possession of, so far the Philippines has adopted a more bureaucratic approach, appointing a mayor for the Spratly Islands. This unlikely representative of democracy has 288 voting constituents scattered across numerous tiny territories.

In 2014 an already jittery situation began to turn into an international crisis. In contravention of international law, China began aggressively to assert control over almost the entire South China Sea. The Chinese navy now patrols the sea as if it were its fiefdom, sailing deep inside the territorial waters of its neighbours and using water cannon to attack fishing boats that cross its path. The Chinese navy also commenced huge island-building projects at Johnson South Reef, Fiery Cross Reef, Gaven Reefs and Cuarteron Reef and, more recently, Subi Reef, Hughes Reef and Mischief Reef. Over the space of a year, the Spratly Islands were transformed into one of the world's most comprehensive military outposts. Harry Harris, commander of the US Pacific Fleet, has nicknamed the result the 'great wall of sand'. Malaysia, Vietnam and Taiwan have also been building up the islets and reefs that they have control of. In each case what is created is a diminutive military base. Vietnam's People's Naval Infantry have been photographed marching across the largest of their possessions, Spratly Island, the eponymous centre of the cluster. This island base barely gives them boot space but, while island-building for peaceful purposes usually requires plenty of acreage, military islands need only be big enough to hold real firepower, which they can do in the shape of a missile silo, landing strip or naval harbour.

Although China is not alone in reshaping the Spratly Islands into military bases, its activities go much further than anyone else's. Jane's Information Group notes that while other nations have 'modified existing land masses', China is 'constructing islands out of reefs that for the most part were under water at high tide'. To build its islands, China has located suitable reefs that can provide stable foundations. Satellite images show how their chosen islands are latched onto by long black snakes that curve through the water, attached to fleets of slug-like boats, many with barges moored alongside. These boats are grinding up the sea floor into manageable particles, then squirting the resulting mix along these black pipes to build up the islands.

These endeavours are led by military considerations. Indeed, the military literally shapes its possessions in its own image. Much like the US's Johnston Atoll, which now looks like a giant aircraft carrier, the new Spratly Islands have come to resemble a fleet of bulky and geometric vessels bristling with weaponry. The before and after photographs of Fiery Cross Reef show the conversion of a natural reef, full of colour and enclosing a large blue lagoon, to a bleached white rectangle with an airfield and gaping square jaw at one end, a military harbour dotted with the black teeth of destroyers and other naval vessels.

International courts have refused to award the Spratly Islands to China, but many countries are scrabbling for evidence that they are theirs. China plays up the idea that the Spratly Islands were discovered in the second century BC by Chinese explorers, and the remains of Chinese coins and pottery found on some of the islands have been seized upon as proof. But these assertions could be replicated by the other claimants, which perhaps explains why China has begun building up more twenty-first-century forms of proof. In 2011 China Mobile announced that residents of the Spratly Islands (of which there were, until recently, none) would henceforth be enjoying full coverage. The inference is clear: the area is just another part of China. The soft power is backed up with hard power. A key exchange occurred on 20 May 2015 when the Chinese navy warned a US military jet away from what it called 'the Chinese military alert zone' over Fiery Cross Reef. This action escalated and widened the conflict; as has China's deployment in February 2016 of surface-to-air missile systems on one of the Paracel Islands, a disputed island group north of the Spratly Islands.

In early 2017, the future US secretary of state, Rex Tillerson, declared that China's island building in the Spratly Islands was like 'Russia's taking of Crimea'. An editorial in the Chinese newspaper *Global Times* commented that 'Tillerson had better bone up on nuclear power strategies if he wants to force a big nuclear power to withdraw from its own territories', and that the US would need to 'wage a large-scale war' if it wanted to push China out of the South China Sea.

The Spratly Islands are the most hotly contested new islands rising from the world's seas, but they are not the only ones. Most are used for peaceful purposes, creating land for new homes and new fields in the Netherlands or Singapore, for airports in Japan or beachside hotels and villas in Dubai. China is also rapidly expanding its civilian island-building programmes both at home and abroad. For example, it is building a harbour island in Sri Lanka. However, the speed and impact of the newly weaponised islands of the South China Sea are turning heads. The Indian government is already seeking to bulk out existing islands into Spratly-type redoubts in the Indian Ocean for its naval forces, and Indonesia and other countries are headed in the same direction. I hope this madness will stop. It does not reflect well on our generation, or our species, that we are malforming once-tranquil islets surrounded by clean blue waters into tank-like bunkers and grey and dusty landing strips.

Bothnia's Rising Islands

Mark Twain's advice, 'Buy land, they're not making it any more', is nowhere less true than in the far north, where Finland and Sweden face each other across nearly 65 miles of frozen water. The provinces of West Bothnia in Sweden and East Bothnia in Finland are separated by the Gulf of Bothnia, and spend much of the year locked in ice. Yet this is one of the world's geological hotspots, so much so that some of it has been classed a World Heritage Site by UNESCO because of its 'outstanding geomorphological attributes'.

A clue as to what is going on is given by the fact that there are small hills a long way inland around here with place names that mean 'island' or 'skerry'. Bothnia is rising up. It is rising so fast that local people have to keep returning to the issue of who, if anyone, owns all the new land.

Around twenty thousand years ago ice sheets covered much of Northern Europe and North America. The weight of all that ice pushed the earth's crust down by around 1,650 feet. Now that most of this ice is gone, the earth is readjusting itself; it is bouncing back. At the present rate, in two thousand years or so the land in Bothnia will have risen so high that the Gulf of Bothnia will close up in the middle, turning its northern arm into a lake. Many of Scandinavia's existing lakes have the same origin. One such is Lake Mälaren, Sweden's third largest, which was also once an arm of the sea. It became a freshwater lake in the twelfth century, and the city of Stockholm was founded on its seaward shore.

It was the Swedish scientist and inveterate measurer, Anders Celsius, better known for his temperature scale, who discovered the phenomenon of geological uplift. In his travels along the Gulf of Bothnia, Celsius had seen how seals rested on rocks of similar height, rocks low enough for them to clamber onto. He also knew that there were written records of the ownership of these rocks, since for seal hunters they were valuable property. Looking back at these records, Celsius found that over the years, once-coveted rocks had been given up and declared worthless because they had become too high in the water. The seals could no longer make use of them and nor, therefore, could the hunters. Celsius carved the date of his investigation, 1731, at the waterline on one of these rocks, to allow future generations to gauge the rate of uplift. What is now called the Celsius Rock is some 105 miles north of Stockholm. Since 1731 it has risen a colossal six feet out of the water. Celsius' carved date is still boldly inscribed, but it is above eye level. This 'seal rock' is so tall that the only animals of any size that can make use of it are birds.

The Gulf of Bothnia has thousands of small islands, but every year new land is seen poking up through the waves. About half a square mile of land is created annually. The Kvarken Archipelago, which lies halfway down the Gulf, is the most spectacular example. This bucolic area, dotted with summer cottages, is in Finland, though it has a large Swedish-speaking minority. Many of the cottages are former waterside fishermen's huts that are now far from the shore. Susanna Ehrs, who has lived here all her life, explained to Public Radio International: 'Where my parents have their summer cottage, where I used to go and swim and we had our boat, is dry land today.' She estimates that 'it's, like, a bit over ten inches in just thirty years, the land uplift. So it's actually visible.'

The rate of rise has meant that the towns and villages in the area have had to keep shifting their harbours. A local tour guide, Roland Wiik, points out other problems: 'The shore-lines constantly change; islands become part of the mainland, and we must keep checking boating routes as treacherous new rocks and islands emerge.' He too recollects a very different landscape: 'Many places where I remember fishing as a child in the 1960s are now dry land covered with trees.'

The Kvarken Archipelago is made up of 6,550 low islands, and counting. After new land first emerges, it takes about fifty years for it to grow large enough and to dry out enough to become usable for house building. Glacial bounceback can lead to arguments about land ownership. To whom does the new land, once raised from the sea, belong? The answer, in Finland at least, is that it belongs to the person who owned the water, not the shore. It follows that if the shore's landowner wishes to build anything on or over the land that has risen, and which now separates his or her property from the water, he or she needs the permission of the person who owned the water from which the land rose. In the Kvarken Archipelago, new examples of the resulting complications have to be resolved every year. Most villages have landownership committees whose job it is to decide how to allocate the extra land. Jan-Erik Mouts, a leader of one such body, explains that in his village they have chosen to sell the new islands to locals, but that in a nearby village they have opted to make the new islands common property. This means that they become 'common land for everyone in the village', and that 'if somebody would like to build a summer cottage' they would have to purchase a long-term rental of the island from the local community.

Some nations pivot on the issue of glacial rebound, such as the UK, where the formerly ice-covered Scotland and the north of England is bouncing back and the south is sliding downward. This seesaw effect is continent-wide. The greatest expanses of ice were centred over Alaska, Canada and Greenland. These parts are rising up, and much of the US is sinking. Morgan DeBoer, who lives near the Alaskan capital of Juneau, showed a *New York Times* reporter round the golf course that he has built on land that was under water when he arrived in the area fifty years ago. Back then, he says, the 'highest tides of the year would come into what is now my driving range area'. The land has risen again since he completed his golf course, so Mr DeBoer is now thinking of adding an extra nine holes.

Uplift creates islands but, in the long run, it leads to their disappearance. As the water drains away, archipelagos are turned into hilly landscapes. Off the coast of Juneau, and today connected to it by a long bridge, is Douglas Island. This island is steadily joining the mainland; the channel between it and Juneau is silting up. One day Douglas Island will be an island no more.

When that day will be is getting harder to judge because of the way global warming and rising sea levels have complicated things. Geological scientists Markku Poutanen and Holger Steffen, who have been looking at the fine balance between these forces in the Kvarken area, conclude that 'from a mathematical standpoint, this is a chaotic phenomenon in which very small changes in the baseline situation have a very big effect on the ultimate situation'.

At the moment it is predicted that the phenomenon of new land creation in the far north will continue but at a slower rate. Despite sea level rises, new land will keep on emerging, year after year, in Bothnia and across much of the planet's highest latitudes. Today these are remote and relatively empty areas. Yet as they rise and the rest of us sink, they are likely to become more attractive to settlers; folk who understand, unlike Mark Twain, that land is, indeed, still being made.

The Philippines' 534 Discoveries

Schoolchildren in the Philippines are taught that their archipelago nation is made up of 7,107 islands. However, in 2016 the country's mapping agency announced that they had found 534 more. They were spotted hiding down south, off the second-largest island of Mindanao. In this tropical, wet region, even small islands are lush; green domes sprouting from a warm sea. Some of the 534 additional islands may have been newly created by the typhoons that plague this part of the world. But most were simply missed. New radar technology has allowed a more comprehensive view.

In the Philippines tiny islands can dot the sea all the way to the horizon, so one has to wonder whether 7,641 is the final number. What, after all, is an island? There are a bewildering number of answers to this simple question. For the Vikings an island was only an island if a ship with its rudder could traverse between it and the mainland. One sneaky claimant of the Kintyre Peninsula in western Scotland is said to have bagged his prize by ordering his men to carry his boat across the land. A much later Scottish definition came in 1861 in the Scottish Census. The Census insisted that somewhere can only be called an island if it is big enough for at least one sheep to graze on it. A distant echo of this definition can be heard from the Filipino cartographers, who tell us that an island has two essential features: some part of it must be above high tide, and it must be able to support either plant and/or animal life. They need to visit their new discoveries in order to provide 'ground validation' of these two points. But neither of these features is particularly watertight. Does every rock jutting up from the sea at high tide count as a separate island? Even less convincing is the second part of the definition, which nods to Article 121 of the United Nations Convention on the Law of the Sea. Here it is stipulated that 'rocks which cannot sustain human habitation or economic life of their own' might be islands, but 'shall have no exclusive economic zone'. Since this extension of national reach is what island hunters tend to be after, it's a crucial distinction.

How many islands 'sustain human habitation' entirely 'on their own'? It's not true of any island I can think of: places like Britain or Malta, for example, are plugged into a global economy and their large populations rely on food imports. And what counts as plant or animal life? Is one sheep enough; or a patch of lichen; two limpets; three squirrels? On this teeming planet, if you want to find plant or animal life you usually can.

Islands are tricky for map-makers. It may seem odd that the Filipinos didn't know how many they had, but this is just as true for many other nations, including far smaller ones. In 2015 the news broke from Estonia that it had 2,355 islands rather than 1,521. As in the Philippines, a more accurate survey was responsible, this time using aerial techniques. It is such a giant leap that one can't help wondering if earlier cartographers had given up when things got too fiddly. Rising land levels might explain some of the new discoveries (see 'Bothnia's Rising Islands'), but most of the blame can be levelled at Soviet-era map-makers. For them, Estonia, tucked in the far north-west of the USSR, must have seemed like a minor outpost, not worth poring over. With Estonia's independence, a patriotically diligent approach began to be taken towards the national map. That the new islands are sources of national pride was driven home by Estonian Public Broadcasting, whose report on the story concluded with the tongue-in-cheek observation that it will 'come as a further blow to neighbouring Latvia, which has a famously low number of islands, officially at one, and that too is man-made'.

Confusion over how many islands there are goes far deeper than faulty or primitive cartography. It is a chronic problem of the map-maker's art. That may sound counter-intuitive: surely islands are easy to spot, and there should be a definite answer to their enumeration? Not at all. If you ask how many islands are in the British Isles (the main islands of which are Britain and Ireland), the number you will get will vary enormously. One recent definition came from a retired marine surveyor called Brian Adams, who suggested that an island is at least half an acre or more. If this is the case, then there are 4,400 (210 of which, Adams says, are inhabited). However, if you go to Wikipedia you'll find that there are 'over six thousand', and that 136 are inhabited. For others the only real answer to the 'how many islands' question is 'innumerable'. But that's hardly satisfactory, and so we struggle on, hoping for an unobtainable sense of completion.

Counting islands is just as big a problem for countries with numerous lakes as it is for those with long shorelines. The number of lakes in Finland is rounded down to 180,000 and the number of islands, many of which sit in these lakes, is given as about 100,000 by many authorities. However, the Finnish national tourist board has plumped for the suspiciously precise figure of 179,584. I can't find any evidence to support such an exact figure, but it is, nevertheless, often repeated, even in academic surveys, as an objective fact.

Nowhere does the desire to arrive at a final answer come to grief more than across the immensities of Russia and Canada. Here, people are prepared to confess that no one knows. You can get a sense as to why from the name of the eastern shore of Georgian Bay, which is just one arm of Lake Huron: Thirty Thousand Islands. That is, very roughly, the number of islands that dot this low-lying, pine-clad part of Ontario. It is the world's largest freshwater archipelago, but it is also a big hint that trying to count all the islands in Canada may be a fool's errand. The national records restrict themselves to more manageable tasks: there are 129 archipelagos and 1,016 named islands, 250 of which are inhabited. One consequence of this plenitude is that Canada has more private islands for sale than any other country in the world. Window-shopping at 'Private Islands Online', I see that 134 are currently available; more than in Europe, Asia and the Pacific combined. Most sell at about US$1 million apiece. Given that they usually lack basic amenities, islands are ludicrously expensive: you can buy a mountain for less. But they tap into something primal; a sense of personal autonomy and specialness that just isn't available anywhere else.

The close association between small islands and big spenders is another reason why the Philippines are so keen to put more of them onto its map. The announcement of the 534 new discoveries stirred a good deal of media commentary and national pride. In the shadow of China's territorial claims across the Pacific Area, especially in the South China Sea (which Manila has taken to calling the West Philippine Sea), the new islands became patriotic conscripts. They were allied with another 'discovery'; namely the Benham Rise, an undersea continental shelf measuring almost 33 million acres reaching out from the country's eastern shore, which the UN decreed in 2012 to be part of the Philippines. This submerged feature tended to get second billing, but in practical terms it is bigger news: it transforms a huge sweep of resource-rich and previously international waters into Filipino territory.

Ironically, some of the islands that attract visitors to the Philippines are not, technically, islands at all. There is the aptly named Vanishing Island, which accommodates visitors in stilt huts since it is entirely covered at high tide. Tourists are also welcome on Kambiling Naked Island, a bare sandbar that also disappears under the waves every day. Our love affair with islands is a form of rapture: we love them even when – perhaps even more when – they are barely there at all. The 534 new islands in the Philippines may be being enlisted as economic and patriotic recruits, but, like all islands, they'll always have their own unique kind of sovereignty.

Wild Strawberries:
Traffic Island

There are many types of love, and many bonds between them. The love of nature and the love of place – biophilia and topophilia – have a particularly intimate relationship. Our lifelong affinity with animals and plants is a passionate commitment that tumbles over and into our bond with place. These love affairs merge in the garden, the age-old site and symbol of human well-being.

This helps explain why we have such a problem with the modern city. Walking or, more likely, driving past barren and stony land – shards of unloved territory in between roads heavy with traffic, or endlessly ripped-up and rubbish-strewn 'development sites' – is an affront, a poke in the eye and, more than that, a source of guilt and loss.

The land should be a garden. It should not just be beautiful; it should be alive. To see others treat it with contempt and, worse, to know that I treat it with contempt – for, of course, I just hurry past, eyes down – is unforgivable. So I am taking some wild strawberries and mint plants to a traffic island, a triangle of land cradled in the cold arms of Newcastle upon Tyne's A167M and A1058. This unlovable and abandoned orphan, around which grinds a perpetual din of traffic and over one corner of which hangs a multi-lane motorway, has been on the edges of my consciousness for a number of years. I went there once before, imagining I could be a motorway Robinson Crusoe; that it would be an adventure and make a good yarn. I hadn't expected it to be quite so murderous and panic-inducing. Feeling so visibly out of place was much tougher and odder than I expected and I scuttled away, beaten.

But now I have to go back; if only because my colleague in the Geography Department, Dr Nick Megoran, an expert in border disputes in ex-Soviet Central Asia, is pressing me to show him this no man's land. He's upped the ante by suggesting we take some university forms with us and do a little light bureaucracy.

It's a fine blue day in late summer, and Nick has just knocked at my office door. He's equipped with two large camping chairs, some sandwiches and a flask of tea. We haven't far to go; it's almost round the corner, but the transition is abrupt. At one crucial point you have to walk on a narrow and uneven cobbled strip into oncoming traffic and then nip quickly across. I've done this before, but it feels very different now that Nick is along for the ride; paranoia is edged out by something approaching a jovial solidarity.

Scrabbling away at the thin earth and hard core, I tuck my strawberries in one corner. They'll be OK, but I'm less confident about the mint. Despite the compost I've brought in my rucksack, as soon as they're in the ground they curl downward, sad victims of avant-garde agriculture. It's at that moment that a policeman turns up. 'Could you tell me what you are doing?' Nick's professional charm is on hand. He explains my eminence in the academic field of doing this sort of thing, and asks the policeman to take our photo for his Facebook page. Given my supposed expertise, I'm rather embarrassed about the sorry state of my mint; a point rubbed in by our new friend. 'Are you sure that's going to grow there?' he asks, in the indulgent tone of someone who's just been told that arsing around on a traffic island is an academic specialism.

The bemused policeman edges his way back to wherever he came from and Nick and I move to the centre of the island, in the shade of some thin saplings, and conduct our business meeting. Nick's a keen Christian, and I am not altogether surprised that the conversation has segued to the Book of Daniel. Several weeks later, after a dose of late-night googling, I realise that I could have waved my hands knowingly and made an allusion to the island as Hortus Conclusus, the 'enclosed garden' that is an important symbol of Eden and chastity in both Islam and Christianity. The following weeks also bring me up close to my new garden in a more conventional fashion, pelting around the surrounding twist of roads at speed in a car. It's hard to catch sight of them and, if anything, the sense of guilt-soaked care is worse than ever. How are my miserable plants going to cope? They were happy in their native soil, and now they're trapped in horticultural hell.

I console myself with the conceit that I'm part of a mighty movement, an urban insurrection of 'guerrilla gardeners'. Indeed, I could have lobbed my strawberries from a moving vehicle now that compostable grenade-shaped seed bombs are on sale, at only £9.37, on Amazon: 'Turn the concrete jungle into a wilderness with our compacted wild flower seed grenades'. One of the doyens of the movement, eco-activist Richard Reynolds, turns up the heat: 'The attacks are happening all around us and on every scale,' he declaims; 'from surreptitious solo missions to spectacular horticultural campaigns by organised and politically charged cells.' This kind of language is heady stuff, and it's making me nervous. I had no idea my strawberries were a form of weaponry. It seems I've amateurishly stumbled into a flower- and vegetable-based war.

Touting the reclamation of waste space for flowers and food as an act of heroic insurrection by armed militants doesn't work for me. After all, it is not passers-by, not the police, not anyone in particular that makes planting a few bulbs feel dangerous, but something more diffuse: it's the modern city, with its impenetrable officialdom and strict spatial routines. The most profound of these routines are the ones that we've internalised; the ones that tell us that anything beyond our front door is none of our business; that the streets, verges and parks – the public realm – are spaces to be passed through and not places to be cared for. When he's describing his own planting activities, which began on the Elephant and Castle Roundabout in London, Richard Reynolds drops the military imagery and allows us to see that 'guerrilla gardeners' are just people who, like so many of us, have a thwarted sense of involvement with the earth around them. His pleasure in reporting that 'the pimped pavements, guerrilla traffic islands and roadside verges are without doubt looking their best ever' is that of the nurturer, not the soldier.

It's an urge as old as life itself, but its modern phase may have begun with Adam Purple's 'Garden of Eden', fashioned on an abandoned lot in Manhattan in the mid-1970s. Instantly recognisable for his long white beard, purple clothes and mirrored sunglasses, Adam's Eden covered 15,000 square feet and was planted in Zen-like concentric circles with tomatoes, asparagus, corn, raspberries and forty-five trees. Adam lost his garden to developers, but even when they don't last long, urban gardens live on in the memory and nearly always attract popular support. Copenhagen's 'Garden in a night' was sustained for several years, despite being created by one thousand people over the course of just one night.

Guerrilla gardening has joined a range of 'street crafts' which attempt to rehumanise public space. Urban crochet, craftivism, yarn bombing, origami and lace graffiti, light sculpture graffiti, miniature installations . . . the labels might not mean a lot to most people, but each of them shares an interest in beautifying and reclaiming boring, unloved bits of the city and turning them into real places, landscapes with stories and imagination. They also share something else: in contrast to the macho vandalism of ordinary graffiti, they dare to bring a sense of care out into the open. Bringing an ethic of care outside may help end the distorted, unhealthy but common idea that kindness and care is something we do in private, something shared only with a very few in our high-walled homes.

That sounds romantic, but it's a risk worth taking. Making it feel easy rather than some act of derring-do will also surely help. I should have learnt that before I set off to my inaccessible traffic island, which is a very lonely place and frankly a little terrifying. It is never likely to be a happy home for strawberries, or anything else. I have moved on to planting sunflower seeds. International Sunflower Guerrilla Gardening Day got me started with this. It's an annual occasion held on 1 May, when people all over the world follow simple instructions: 'In open ground, ideally not too compacted (loosen with a small hand tool first), plant a single seed about 2cm deep and cover with soil. Water if necessary, but in Northern European climates, usually the planting is all that's necessary. Suitable locations include an infrequently tended municipal flower bed, pavement verge or nature strip.' It's odd, fascinating, how this simple act can make me feel so nervous. Am I allowed? A better question is, will I allow myself? It is, after all, the most natural thing in the world.

II.
Enclaves and
Uncertain Nations

This journey starts in one of the most important yet often-ignored type of enclave, the language enclave. In the Ladin Valleys of Northern Italy, an ancient tongue is struggling for survival. But what is Ladin? It turns out that enclaved languages often break up into numerous variants. The same is true of enclaved religions, like the Jewish 'eruv'. The example I visit here is more unusual than most: a zone of strict orthodoxy, fringed by the swaying palms of Bondi Beach. Enclaves are usually thought of in national terms, and the following chapters explore a diverse collation of geopolitical enclaves and uncertain nations. At the heart of Asia, the bountiful Ferghana Valley has more enclaves than any other part of the world; diminutive yet increasingly viperous nests of border-making. Hostility and conflict are also to the fore in Morocco's 'Saharan Sand Wall', a stupendous barricade that divides and conquers the nascent nation of Western Sahara. There are no walls yet around that part of the Ukraine rather high-handedly labelled 'New Russia' by pro-Russian groups, but, as we shall see, the separatists appear as concerned with rekindling memories of Russia's colonial past as they do with fashioning a new nation. All these fractious entities might benefit from a little spiritual guidance from what is, perhaps, the smallest and certainly one of the most difficult-to-classify nations in the world, the Sovereign Military Order of Malta. I finish with another micronation, albeit a far more ephemeral one. Post-Brexit, my street in Newcastle declared itself 'The Stratford Republic'. After all, in an era of fragmentation and breakaway statelets, why shouldn't we join in?

The Ladin
Valleys

You can be in the middle of a language enclave and not know it. That was my experience, anyway. I was staying for a week in the Zoldano Valley, which tumbles down among the fantastically spiky peaks of the Italian Dolomites. The fields in this part of the Alps are hemmed with growing armies of holiday homes, and it was out of one of these fat-timbered, pseudo-homesteads that I strolled off each morning, up into the high summer meadows. I had no idea I was in one of Europe's most famous language enclaves, home to one of the many dialects of the Ladin tongue.

A few language enclaves shout for your attention. In Ireland, border signs announce 'An Ghaeltacht', making sure you can't miss the fact that you're entering Irish-speaking territory. But this is more the exception than the rule: languages rarely have visible borders, and those parts of the population that speak a minority tongue tend to do so only among themselves.

I'm a little ashamed that I spent my days panting for breath in what proved to be very high and daunting terrain, and barely registered the intricate linguistic geography of this alpine borderland. To ease my guilt, on returning home I began to listen to recordings of Ladin. To my cloth ears, it sounds like an Italian speaker switching back and forth into either French or German but with a unique vocabulary. Different speakers from the different valleys sound either less or more German, less or more Italian or less or more French. Linguists hear something else, including lots of influences from neighbouring dialects as well as ancient Latin (as the name implies, 'Ladin' is, in part, a preserved form of Latin).

Although only written down in the nineteenth century, Ladin has a wide-ranging oral folklore. Its cultural jewel is the Ladin national saga, the *Fanes' Saga*. It is an epic ballad of betrayed kings, battling dwarfs and secret alliances with marmots. Marmots were not a random choice. On my alpine hikes I saw plenty of these tubby and very fluffy burrowers, nervously dashing between their tunnels. Surely I could master a few Ladin phrases to greet them? Like *bun dé* (hello), *giulan* (thanks), and *co aste pa inom?* (what's your name?). These might work on marmots but it turns out that learners of Ladin do not have an easy task. The language is like a delta; distinct rivulets run along every valley. There are six major forms of Ladin, each of which breaks down into a variety of dialects. The Zoldano Valley form of 'what's your name?' is *ke asto gnóm?*

The total number of speakers of all these forms and dialects of Ladin is a meagre 33,000. But even totting up the number of Ladin speakers is not straightforward. The number could be swollen if you included other local languages which appear to be country cousins of Ladin, namely Nones, as spoken in the Non Valley, and Solandro as found in the Sole Valley. Then there are Ladin's more distant relations, such as Friulian, sometimes called Eastern Ladin ('what's your name' here is *ce nòn àstu?*), which is spoken in the far north-eastern parts of Italy, and Romansch (*co has ti num?*), one of Switzerland's four official languages.

I'm keeping things simple. I'm only mentioning one language minority, Ladin, and its offshoots. I don't dare foray further, into the varieties of German on offer in the Dolomites. German is the majority language in South Tyrol, the Italian province just north of the Zoldano Valley, and there are a number of languages derived from German, such as Mócheno and Cimbrian, that are spoken in very local pockets across the wider region. Each, inevitably, has a variety of dialects. The same is true of Ladin's relations. Friulian has central, north-eastern, western and northern dialects, while Romansch can be divided up into Sursilvan, Sutsilvan, Surmiran, Putèr and Vallader, each with its own written form.

My good intentions are starting to wane. The level of differentiation and tiny scale of these language pools has a curious effect on me; it's making me anxious. As soon as I think I've got things clear, they recede and break up. I guess my anxiety is a reflection of my monolingual background. I'm just not used to this kind of mosaic. I'm surprised and disappointed with myself. I grew up in London, where people have become rather boastful about the city's multiculturalism. Yet the diversity of this beautiful rural backwater is turning out to be too much for me. My anxiety is certainly out of keeping with the celebratory and relaxed disposition of the locals. People here take enormous pride in their linguistic localisms. There is a kind of 'enclave culture' at work, highly attuned to the micro-scale and the pleasures of staking a claim on each community's linguistic uniqueness. But as an outsider, it can feel as if these lists of languages and dialects might go on forever. There are just so many of them. I'm starting to feel nostalgic for the high meadows. I was happy up there, with my marmot friends: it's the human world that I struggle with.

It turns out that if I'm seeking enclaved languages that do not immediately melt into myriad streams, then I am probably looking in the wrong place. The borders of the 'Old World' are piled high with history. Combine that with a mountainous crossroads region and you have a recipe for daunting plurality.

And these Italian mountains are just a beginner's slope. If things appear complex here, then you'd better avert your gaze from the Caucasus Mountains in Southern Russia. There the variety of tongues is so dense that it is unclear if all the languages have even yet been identified. The same can be said of the island of New Guinea, just north of Australia. It's another region of old cultures and deep valleys. It is sometimes reported that there are 750 languages spoken in New Guinea, but others say there are 850 . . . perhaps.

In the New World, where one can find 'old tongues' transplanted and maintained, things are a bit more manageable.

Many minority European languages have been preserved and hybridised in the Americas, like 'Patagonian Welsh', which is spoken by a few thousand descendants of Welsh miners in the south of Argentina. Another example is the splendidly named 'Riograndenser Hunsrückisch', an archaic German dialect spoken by the distant offspring of German migrants in Brazil and which contains both Portuguese and Amazonian elements. Another German anachronism found in Brazil is Germanic Pomeranian, an Old World dialect little used now in Europe but still spoken in pockets of Brazil.

The liveliest spots in which to find language enclaves in the Americas are the big cities. Language hunters have a soft spot for New York, which is home to numerous immigrant tongues that have died off, or are in danger of doing so, back home. One example is Garifuna, which is derived from the indigenous languages of the Caribbean and is spoken by descendants of those Africans who became fully acculturated into native Carib culture. Today Garifuna is more likely to be heard in the Bronx than anywhere near the Caribbean. Similarly, you don't hear much Vlashki in Croatia anymore. Unlike Croatian, which is a Slavic language, Vlashki is a Romance language, coming to Croatia by way of shepherds who migrated from Romania. There are only a few hundred people who still speak it back home, but Vlashki speakers in New York are trying to save the language, and have created a digital archive.

Ladin doesn't seem to have made it to New York, though a venerable tongue it is often confused with – Ladino, a language that combines medieval Spanish and Hebrew – hangs on in ageing Jewish communities in Brooklyn. Ladin doesn't need assistance from across the sea. It has the kind of help and protection in Italy that is a distant dream for most minority languages. There are Ladin cultural centres, Ladin radio stations and Ladin TV and, in some provinces, a linguistic quota system for government positions and public service jobs. From being on the edge of extinction, Ladin is now affirmed and respected. But even in this supportive environment there is an attenuation process at work. Keeping Ladin going is ever more an act of will, a conscious effort. It has slid from being an ordinary thing, shouted out by kids running from the school gates, into becoming the preserve of well-intentioned activists and committees. It's not going to disappear completely any time soon, but it is on the way to becoming a museum piece. One of many in the charming but increasingly frozen linguistic kaleidoscope of the Dolomites.

The Eruv at
Bondi Beach

An eruv is a religious enclave that is commonly formed by stringing up a piece of wire between poles which symbolically turns a public space into a private one. Many very orthodox Jews believe that on the Sabbath day they are forbidden to carry anything, irrespective of its size or purpose, within a public area. That means you can't do much: you can't push a pram or carry a book. The eruv gets round that problem.

It's hard to say how many eruvin there are in the world, as they range in scale from a small patch in front of a home to a whole city (Boston is contained by a single eruv). Most are in Israel and the US, but there are a growing number elsewhere. Many eruvin incorporate pre-existing features of the landscape. The eruv that is centred on Bondi Beach in Sydney is created from a combination of the sea edge along

with wires and fencing that have been approved by the local eruv committee.

The overlaying of two such dissimilar landscapes – a famously hedonistic surf-and-sun lifestyle and dutiful religious orthodoxy – makes Bondi an intriguing example of an eruv, especially since it is unknown and invisible to so many passers-by. The idea of being in a place without knowing it is one of the unforeseen consequences of multiculturalism. Different ethnic groups often have different perceptions of which geographical boundaries and identities matter. It can be a disconcerting discovery. A few years after the Bondi eruv was created in 2002, I rented a single room for a few weeks in the university district of Randwick. I was hoping to get into a regime of jogging along the seafront, but it didn't work out. It wasn't the Jewish orthodoxy that put me off, but the far more dominant culture of beach-beautiful youth. Intimidated by the succession of confident, sculpted bodies that sped past me, I ended up walking; several times a week, up past a romantic clifftop cemetery and on towards Bondi Beach. Not until my last week, when it was mentioned to me in passing, did I realise that at Hewlett Street I was entering sacred space.

My first question was, what's an eruv? Outside of Israel, these are places that mean a lot to those who make use of them, but to almost everyone else they are unseen. The weightless, indiscernible quality of the typical eruv is, in part, a reflection of the unique kind of wall that surrounds it. The eruv should, in theory, be bounded by a solid wall. The fact that the boundaries of most eruvin don't look anything like a normal wall is explained by the rabbinical argument that walls can have any number of windows and doorways and, hence, a wall does not have to be solid. The eruv is, in effect, a vertical plane of windows and doors: a transparent wall that is physically but not religiously permeable. In fact, the Bondi eruv is sometimes heralded as 'one of the few eruvin in the world that has a majority of solid boundaries'. This makes it sound much more substantial than it is. These 'solid boundaries' are, if anything, even less noticeable than the wholly wired ones, since they are mostly made up of pre-existing and otherwise normal-looking fences and cliffs.

A detailed map of the Bondi eruv has been produced by the local eruv committee, and it shows how it dips back and forth along the coast, taking in thirteen Jewish synagogues and community centres. The eruv is about rules: rules about permissible work and permissible space. Users of the Bondi eruv are warned that: 'Only the car park level above the promenade at Bondi Beach is wholly within the eruv boundary. It is not possible to walk the length of this promenade and stay within the eruv'. And further: 'If unsure about the boundary, please consult your Rabbi'. Such finely grained demarcations are designed to allow necessary activities while ensuring a sense of religious sanction. Eruvin encourage regimes of micro-geographical self-surveillance, as illustrated by this rabbinical instruction to the users of one New York eruv:

> The hospital is included in the eruv. However, when walking on Oceanside Road to the hospital, just at the entranceway into the parking lot, there is a telephone pole. That is the Boundary. You must be very careful to

walk between the telephone pole and the hedges of the parking lot. DO NOT WALK outside of the poles or the opposite side of Oceanside Road, as you will be outside of the eruv boundaries.

The Bondi eruv was set up after a couple of inspections from one of the world's leading authorities, Rabbi Shimon Eider of Lakewood, New Jersey. He fully approved of the eruv, declaring that it was 'one of the largest and most beautiful eruvin in the world'. Being eruv-compliant is a complex business. Rabbi Eider was required to inspect the coast from a boat to make sure the small cliffs satisfied the specific regulations concerning the size of rises and drops that can be used to demarcate 'private' space.

Thankfully the Bondi eruv does not overlap with any others. Two geography professors who have studied the growth of eruvin, Peter Vincent and Barney Warf, write that 'rabbis frequently cannot agree on acceptable boundaries from a strictly Talmudic point of view'. Where overlaps have happened, as they have, for example, in Brooklyn, the confused faithful have been assisted by the practice of attaching colour-coded ribbons to each eruv's wires.

It is not just the issue of drawing the borders that is complex; there is considerable debate on what is and is not permitted on the Sabbath within and outside of eruvin. One set of guidelines I read starts with the statement that the 'wealth of details provides for a lifetime of scholarship'; which is no doubt true but hardly encouraging. It turns out that carrying things may be permissible in an eruv, but that doesn't mean you can carry anything. You can't, for example, carry objects that you are forbidden to use on the Sabbath. This covers umbrellas and pens and other items that are designated as 'muktzeh'. At this point, not having a 'lifetime of scholarship' to draw on, I had better defer to the exact words of at least one of the many religious authorities who crowd this topic with their learning. 'Many objects have been designated by our sages as muktzeh – we are forbidden from moving them'. There are numerous categories of muktzeh; to mention just one: 'Objects which have no designated use, e.g.: stones, plants, flowers in a vase, raw food (inedible in its present state, such as beans); an object that has broken and become no longer useful, such as a broken bowl, a button that falls off'. But a distinction is then drawn: it is forbidden to move these things

> directly with one's hand or even indirectly with an object (such as sweeping it away with a broom). However, muktzeh may be moved in a very awkward, unusual manner, with other parts of the body, e.g.: with one's teeth or elbow, or by blowing on it.

If I returned to Randwick today I'd find that my modest room in Sydney has also been designated as part of an eruv. The new Coogee-Maroubra Eruv was trialled in 2016 and extends across much of Randwick, being 'built to assist and enhance the Shabbos [Sabbath] experience especially for parents with young children and toddlers in strollers'. Trying to explain why eruvin have 'multiplied rapidly recently', Peter Vincent and Barney Warf tell us it's a reflection of what they call 'resistance and withdrawal from the hegemony of commodity fetishism'. Professors of human geography have developed a weakness for this kind of explanation: one can hardly take one's hat off without being co-opted into resistance against capitalism. A more obvious and plausible explanation for the spread of eruvin is that we have seen an upsurge in religious conservatism and literalism across the world and that Judaism is no less affected by this than any other faith. It would be nice to imagine that the rise of eruvin is not part of the religious and identity-based politics of our time, but it would not be convincing. I suspect that, as they spread, eruvin will become more visible and contested components of the multilayered urban ethnic landscape.

The Ferghana
Valley

The Ferghana Valley is at the centre of Asia, nearly 1,250 miles from the nearest ocean; the middle of the middle. It lies in the very heart of 'Central Asia', a rather nebulous but very central-sounding label that also puts it at the crossroads of three former Soviet states: Kyrgyzstan, Uzbekistan and Tajikistan.

It is a riven heartland. The Ferghana Valley contains eight fractious enclaves. There are two marooned outposts of Tajikistan and four bits of Uzbekistan in that part of the valley that is in Kyrgyzstan, as well as one Tajik and one Kyrgyz territory in the part that is Uzbek. Although they appear as solid lines on most maps, many of the borders hereabouts are actually far from certain and very much disputed. Over recent years the valley's many national boundaries have been witness to a lot of what the outside world usually classes as 'ethnic violence'.

Not that the outside world bothers much to look. Perhaps the name 'Ferghana Valley' is to blame. It encourages us to think that this is an insignificant place, bizarrely complex and prey to merely local bouts of bother. In fact, fourteen million people live here, and the 'valley' is a vast, low-lying plain bigger than Israel. It was once legendary for its abundance: from the high, icy mountains that surround the valley, melt-water tumbles down to nourish rich acres of apricot, peach and walnut orchards as well as the region's famous mulberry trees. It is these orchards, this fertile land, which are at the root of much of Ferghana's ceaseless border-making and border-taking. Today the waters are faltering, the valley is becoming more arid and population pressure means that good land is getting scarcer. A life-and-death struggle for resources has begun; a struggle which is played out in the language of ethnic and national identity and hostility.

The Ferghana Valley is a microcosm of a much wider phenomenon; namely the way environmental crises are turning into political conflicts. The violence in the valley is episodic. To take a typical example, on the morning of 11 January 2014, Tajik and Kyrgyz forces began shooting at each other around the village of Vorukh, a Tajik exclave that sits deep inside Kyrgyzstan. Accounts differ about who started it, but it seems it lasted about an hour and that several border guards were injured. What is clearer is that the shooting was sparked by mutual suspicion that the other party was attempting a land grab over a 50-square-mile area of fertile land. A tense stand-off ensued, as it has in many other locations across the valley; a stand-off which, punctuated by more bouts of gunfire, is still going on.

If the borders were clear-cut and everybody was happy with them, these flare-ups wouldn't occur. But no agreed political map of the area exists. Locally people call it 'the chessboard border' because of its complex patchwork of sovereignties. Villagers who wish to access their local water supply or market have to cross and recross international borders, a process that is made even more fraught by having to deal with corrupt officials. The region has seen a proliferation of badly paid and poorly trained border guards.

The different sides use very different maps to bolster their claims. The Uzbeks point to maps drawn up between 1955–9, but the Tajiks favour the ones that favour them, dating from 1924–7. Back then, it hardly mattered. Each of the three countries with territory in the valley were Soviet Socialist Republics and fraternal allies within the Soviet Union. Some historians have begun using the phrase 'multicultural empire' to describe the former USSR, and it captures an important truth. Far from being monolithic, the USSR knitted together an incredible diversity of religions, languages and cultures. Moscow created numerous new nations: all the ex-Soviet countries ending in 'stan' ('stan' is an ancient Farsi word for country or land) were its inventions. Kremlin politicians continually tinkered with their borders, shoring up support with territorial gifts; goodies which would often be subsequently altered or withdrawn. This erratic approach helps explain a lot of the 'wigglyness' of the borders today. Although Russia dominated the Soviet 'multicultural empire', the USSR allowed and encouraged ethnic mixing, with people freely travelling, intermarrying and speaking Russian as a shared lingua franca. When, in the early

1990s, the outer edges detached themselves from this union of peoples, a lot of that mixing and interaction ground to a halt and there commenced a rapid shift towards national border-making and ethnic purism.

The new generation of border guards no longer all speak Russian, so don't have a shared language. Uzbekistan moved from a Cyrillic to a Latin alphabet in 1995, while Kyrgyzstan and Tajikistan did not; so printed matter, including highway signs, on either side of the borders now looks very different. Moreover, only some of these countries have prospered, so economic gaps have grown. Uzbekistan is relatively wealthy and the roads on its side of the border are fairly well maintained. The Uzbeks have grown rather dismissive of their neighbours. In 1999, the country's president, Islam Karimov, explaining why a cross-border bus service, along with many other routes into the Ferghana Valley, had been suspended, commented: 'Kyrgyzstan is a poor country, and it is not my job to look after their people'. Using a Soviet-era zero-sum logic, he explained that if 'every day five thousand people' come into his country from Kyrgyzstan and 'if each of them buys a loaf of bread, there will not be enough left for my people'.

The creation of ethnically defined nations means that many citizens within those nations who happen to be of the 'wrong ethnicity' do not feel wanted. This helps explain why the enclaves are not necessarily loved by their own nations. The consequences of Soviet tinkering mean that they are not always occupied by people of the same heritage as the 'mainland'. The enclave of Sokh, one of the largest in the region, is one such orphan. Ninety-nine per cent of the population in

Sokh are Tajiks, but it is surrounded by Kyrgyzstan and is the sovereign territory of Uzbekistan. The Uzbek government has been indifferent to the plight of this distant outpost, especially so since the Islamic Movement of Uzbekistan (IMU) has tried to gain access to Uzbekistan through Sokh. It is now cast as a security problem, with the border guards as much there to monitor the activities of the people within the enclave as to stop it being encroached upon by outsiders. Uzbek authorities have laid the enclave's border with anti-personnel mines in an attempt to prevent the Islamist incursions. My colleague, the political geographer Nick Megoran (see the chapter 'Wild Strawberries: Traffic Island'), has interviewed some of those affected, including Aybek, a shepherd boy from Sokh wounded by a landmine in 2002, who told him: 'There were no warning signs put up before then – afterwards they put them up, but they still didn't give me any compensation.'

All of these new identities and borders are given an edge and turned into sites of desperate conflict by the environmental crisis that overshadows the region. The glaciers and snows of the range that the Chinese call the 'Heavenly Mountains' at the far end of the valley have long been key in transforming the area into an agricultural oasis. But recent studies have shown that, between 1961 and 2012, the Heavenly Mountains have lost 27 per cent of their ice. Dr Daniel Farinotti, who led the research, explains that this means the glaciers have lost 'each year as much water as all the people of Switzerland, including industry' get through in six years. It's not only the enclaves that are feeling the resulting water stress. Speaking to Al Jazeera, a local community official from a small town inside the Uzbek part of the valley put the matter in blunt terms: people, he said, are 'ready to kill each other for water'. Farmers in his town are getting into fist fights about the timing and allocation of piped water.

In the absence of cooperation, the favoured solution to the problem of the enclaves has been to create a bypass infrastructure. So, for example, to stop their drivers getting hassled at the border, the Kyrgyz are trying to build a road round the disputed enclave of Vorukh. However, the Tajiks say that the road is going over their land, and further disputes and conflicts have ensued. People in these hillsides can arm themselves quickly, with shotguns, Molotov cocktails and stones, and contractors need a high level of protection.

In the middle of the map, the rich soils of the Ferghana Valley should be a Garden of Eden and a place of solace and calm. Instead, the heart of Asia is a sour tangle of rivalries. Many of those who live in the valley look out over the bitter results with nostalgia for yesterday's 'multicultural empire'. If the cost of national sovereignty is endless conflict and a failure to tackle environmental problems, then it is surely too high a price to pay.

The Saharan
Sand Wall

This is a story about an unrecognised border and a forgotten war. The Saharan Sand Wall is the longest active military barrier in the world. At 1,367 miles, it is longer than the distance between London and Saint Petersburg. On its eastern side is land under the control of the Sahrawi Arab Democratic Republic, a country that has full member status at the African Union and that has been recognised, at one time or another, by nearly half of the world's governments.

The Sand Wall is often called the Saharan Berm, a 'berm' being a raised bank. From the air it resembles ancient archaeology. It's easy to see it on Google Earth, snaking raggedly down one side of Western Sahara before looping westwards to the Atlantic. It takes a while to realise that it's not a remnant of the distant past. With its sand-blown ridges and intermittent sandbanked forts, it looks like a relic from the Roman period, a lost rim of empire now reclaimed by its desolate surroundings. But the wall turns rather too many abrupt corners; and some of those forts appear to have airstrips. It's not ancient but modern; guarded by 90,000 troops, the surrounding land is some of the most dangerous in the world, riddled with an estimated seven million landmines. Tunnels have been dug for added protection for the troops and for relief from the sun. In many places the wall consists of two sand berms some 6 to 9 feet tall, and a ditch on the eastern side, which run across the parched desert like fingernails dragged down the map.

The wall was started in 1980 and was constructed in six sections from north to south, until it was completed in 1987. There are five 'breaches', or openings in the wall, which appear to be there to allow the Moroccan army to pursue the forces of its enemy – the Sahrawi Arab Democratic Republic – better known as the Polisario Front. Polisario's claimed territory is the whole of Western Sahara, but most of that is firmly behind the wall and under the control of the Kingdom of Morocco. The would-be Republic is left with little more than barren desert. To make matters worse, many of the Sahrawi Arab Democratic Republic's supposed international allies have turned out to be fair-weather friends. After a spurt of popularity in the 1970s, many countries have frozen their recognition of the Republic, increasingly unconvinced that Western Sahara, which has a population of only 600,000, will ever make it as a state.

Ruled from Spain between 1884 and 1975, today the Western Sahara is Morocco's big backyard. It is about 60 per cent of the size of Morocco, and for years it has been both walled-in and colonised. State-sponsored schemes encourage

Moroccans to settle there. The 1975 Green March saw hundreds of thousands of patriots make the long journey, many on foot and carrying Moroccan flags, pictures of the king and copies of the Koran. Although claimed as a peaceful reclamation of empty lands, tens of thousands of the people who lived there, the Sahrawis, were forced into exile. By some estimates, Moroccan settlers now make up at least two-thirds of the inhabitants of Western Sahara.

But the wall signifies weakness as well as strength. It was built because the Moroccan military was being outpaced by Polisario in the 1970s. Despite their superiority in numbers, they were being defeated. The wall turned that around, containing and immobilising Polisario and forcing them into a ceasefire that has lasted a quarter of a century. It may look old-fashioned, but it has worked. This very old technology is brutal, crude even, but it is effective. The wall has also played a big part in the longevity of the conflict. It has frozen the two sides into their respective camps, ensuring Polisario cannot enter Western Sahara, but also creating generations of resentful, displaced Sahrawi who vow to return. It has also turned Morocco into a militarised kingdom which spends vast sums on defending its extended realm.

Although the Sand Wall is sometimes called the Berlin Wall of the Desert, it is a unique creation, especially so for being in a part of the world where remoteness, low population density and some of the hottest temperatures on earth mean that most borders are unmarked on the ground. The fact that part of the wall dips down into Mauritanian territory reflects just how hazy and unpoliced borders can be in the Sahara.

'Sahrawi' is Arabic for 'Inhabitant of the Desert'. The word covers a mixed group of tribes, many of whom now live in refugee camps in Algeria and Mauritania. For its opponents, the Sand Wall is a way of stopping these people returning home and an act of colonial force by the Moroccans. In their temporary capital of Tifariti, an oasis town located east of the wall, the Sahrawi National Council, Polisario's government in exile, tries to exercise authority. One of their main tasks is to tell the world they still exist. Their story was never well known, but today it is in danger of being forgotten altogether. It was not until the 1930s that Spain was able to gain effective control over this land. In the 1950s, phosphate was discovered in Western Sahara and the Spanish invested in infrastructure, including schools and cheap housing close to the mines. It appears to have been these new opportunities which led to many Saharawis giving up their nomadic lifestyles, settling in towns and turning to nationalist politics. Thus the idea of a Sahrawi nation began to emerge. The Sahrawi have been claiming independence since 1975, the year that Spain left what used to be called 'Spanish Sahara', leaving the door open for neighbouring countries to rush in and stake their claims. The Moroccan government likes to label Western Sahara as their 'Southern Provinces', or Greater Morocco. The official position is elaborated by a professor of law called Said Saddiki, who explains that the wall is a defensive structure and dismisses the viability of a Sahrawi 'microstate'. Saddiki talks the kind of language that the international community listens to. In a region flooded with weapons from Libya and prey to Islamist violence, the Sand Wall has been rebranded as a counterterrorist device. The Sahara Wall, Saddiki says, is 'an effective impregnable obstacle to the movements of Islamic military groups', and its existence 'explains why Western Sahara remains away from the attacks of these groups in comparison with other regions'. Saddiki predicts that if Western Sahara ever got independence it would quickly become a 'failed state', and be one of Africa's 'hotbeds for military groups, arms and drug trafficking'.

Professor Saddiki has another card to play too: the wall, he says, slows the flow of migrants getting into Europe, those 'irregular sub-Saharan migrants' who would otherwise stream up through Morocco. It is ironic, then, that one of the less advertised uses of the wall is Morocco's practice of dumping illegal migrants on its eastern side. Migrants in the north of Morocco from other African countries, as well as from South Asia, have been driven south and forced out into the desert. Lost and dying of thirst, some have been rescued by the Polisario Front.

The political temperature is heating up around the Berlin Wall of the Desert. Many refugees have grown tired of living in limbo. Senior Sahrawi politicians are publicly regretting the decision to call off the armed insurgency. Meanwhile Morocco is getting more confident. In 2016 it built a major road into the disputed territory, and plans for the integration of 'Greater Morocco' are proceeding apace. In his speech to the nation on the fortieth anniversary of the 'glorious Green March', King Mohammed VI talked about new road and rail routes into what his government calls the 'Southern Regions', as well as ports and hub airports. 'Following the epic achievement of liberating our land and shoring up peace and security,' the king informed his subjects, 'our country has sought to enable the Saharan populations to become full-fledged citizens and enjoy a dignified life.'

New Russia

It was minus 20 degrees outside, but I was in a warm radio studio in Moscow waiting for what I thought would be a politely humorous chat about micronations. I was disconcerted to find that the presenter was on his feet and angry, an angular young man bouncing on his heels and in no mood for small talk. His opening salvo was 'What about Donestk? What about the victims in the Donbass; where is their micronation?' I was being rapidly educated about the fact that small enclaves and little nations are not a quaint topic in this part of the world, but a very raw and very urgent matter. I blustered and waffled; the presenter's eyes narrowed into something that, I hoped, was pity and I silently promised myself to learn more about this disputed borderland.

New Russia, or Novorossiya, was first established in the mid-eighteenth century in what is now Southern Ukraine. It was a Russian equivalent to the claims other European powers were making on virgin territory such as New England or New Spain.

But unlike settlements in the New World, which saw native people pushed out of the picture, New Russia was always a site of many voices and complex ethnic politics. Over recent years it has been reborn, as a state that vanished almost as soon it was created but also as a stubborn ideal, a key part of Europe's new politics of ethnic separatism. The story of Novorossiya takes us to the heart of one of the most contentious of the twenty-first century's many fragmented landscapes.

The words 'New Russia' send a chill down the spines of many Ukrainians. It sounds like a land grab. Just two little words, but they hand half of Ukraine to Russia. Conversely, the majority of the Russian public sees Ukraine's Russian minority as victims of anti-Russian hostility; and a small but dedicated group want to see the return of Russia's old colony.

Novorossiya was founded in 1764 and existed up to 1918, when it was parcelled up into the new Ukrainian Soviet Republic. The lands of New Russia were handed out to a mix of settlers and they formed a lively multicultural community, with Ukrainians and Russians joined by Serbs, Jews, Romanians, Greeks, Italians and many others besides. People came from all over Europe to settle here, and some even lent their names to the newly founded towns. The coal mining city of Donetsk, the scene of so much separatist violence in recent years, was founded by a Welsh immigrant called John Hughes and used to be called 'Yuzovka' (or 'Hughes-town'; 'Yuz' is a Russian transliteration of 'Hughes').

Novorossiya was never meant to be an ethnic enclave for any one group. However, large-scale Russian migration and the suppression of the Ukrainian culture and language by

the imperial government meant that, by the end of the nineteenth century, all of Ukraine's cities were Russian-speaking,
and Ukrainians who wanted to get ahead had to learn Russian.
Although 'Russification' was widespread, the industrial towns
of the Donbass, the eastern province that takes in the regions
and cities of Donetsk and Luhansk, were particular centres
of Russian settlement. The Donbass became one of the
powerhouses of the USSR. In 1987 it produced 26 per cent
of the entire country's coal. It was also one of the bastions
of communism, both before the fall of the USSR and afterwards. It was no great surprise that a 1994 referendum in
Donetsk and Luhansk showed that 90 per cent of the population supported giving the Russian language equivalent status
to Ukrainian. The Kiev government refused to countenance
this petition.

In late 2013, a period of ethnic violence and turbulence
began in Ukraine, a period which has not yet ended. The
'Ukrainian revolution' started in 'Independence Square', a
'revolution' that has become known as the Euromaidan (or
'Euro Square') Revolution. The name flags the political direction: it was a movement of Ukrainians who saw their future in
Europe, not Russia. These protests succeeded in removing the
Russian-backed president, but they did not speak for a sizeable
pro-Russian minority of the population. Anti-Maidan protests
swept cities in the south and east. It was at this moment that
Russia stepped in and took over the Crimean Peninsula,
always the most Russian of Ukraine's regions. In other parts
of Southern and Eastern Ukraine, the pro-Russian movement
also took on an insurgent and military character.

The ghost of New Russia was stalking the country. On 7 April 2014 fighters occupying government buildings in Donetsk declared the creation of an independent 'Donetsk People's Republic'. A series of armed offensives and counter-offensives between pro-Russian separatists and the Ukrainian army began. The Ukrainian army's 'anti-terror' campaigns did little to stem the spreading power of the Donbass militants, and on 4 May the flag of the Donetsk People's Republic was raised over the police headquarters in Donetsk. Other towns and cities soon fell under separatist control. Although the separatists do not control much of the countryside, they hold the major cities. About 50 per cent of the population of the Donetsk region – around 1,870,000 people – lives under rebel rule.

The war of the Donbass had started. It has been a grinding, low-intensity war of attrition, causing thousands of deaths. The presence of Russian paramilitaries and Russian backing explains why the Ukrainians could not quell the uprising. On paper the Ukrainians have the larger and better-equipped force, but the separatists have been continuously resupplied from across the border. All sorts of Russians, ordinary citizens as well as soldiers, made their way to fight for the cause. Public recruitment rallies seeking willing men to fight in the Donbass were held in many Russian cities. However, pro-Russian volunteers also came from further afield. They included Chechen fighters and so-called 'new Cossacks', who proclaim it a sacred duty to retake Russian lands and defend the Russian Orthodox Church.

A week or so after the Donetsk People's Republic declared

independence, the Russian president Vladimir Putin broadcast some telling remarks. 'I would like to remind you,' he told his radio audience, 'that what was called Novorossiya back in the tsarist days – Kharkov, Luhansk, Donetsk, Kherson, Nikolayev and Odessa – were not part of Ukraine back then.' Warming to his theme, Putin then began musing on the vagaries of history: 'These territories were given to Ukraine in the 1920s by the Soviet government. Why? Who knows? They were won by Potemkin and Catherine the Great in a series of well-known wars.'

This open-ended querying of the political status of Ukraine's southern half set alarm bells ringing in Kiev. A few days after Putin's historical speculations were broadcast, the second of the eastern Donbass regions, Luhansk, was declared to be an independent 'People's Republic'. Keen to seize the ultimate prize without delay, the leaders of both new 'People's Republics' proclaimed their intent to federate as Novorossiya. The newly animated New Russia always seemed more an aspiration than a reality, even after it was called into being. Another new entity, the New Russia Party announced its existence on 13 May 2014. Chairman of the New Russia Parliament, Oleg Tsarev, proclaimed that New Russia was 'ready to absorb all the other republics of the South-East of Ukraine and, it is possible, in the West too'. Pavel Gubarev, the 'People's Governor' of Donetsk, declared that the New Russia Party, the new state's political voice, will be 'led only by those people who in this difficult time showed themselves as true patriots of their Motherland and proved themselves as true fighters and defenders of their Fatherland'.

The new state called for 'national elections' and a competition for a national anthem was announced. But the fragmented
and conflict-ridden nature of the territory it claimed drained
the new state of credibility. The war of the Donbass ebbed
and flowed; ceasefires came and went – eleven in total –
and, increasingly, outside observers began to refer to it as a
stalemate. New Russia limped on for several months, and in
December 2014, it issued a renewed claim to sovereignty, this
time proclaiming itself to be a direct descendant of the USSR.
On 1 January 2015, New Russia's leadership threw in the
towel and announced that the nation had been 'suspended'.
However, many regard it not as dead but, in the words of one
separatist politician, 'moving to another plane', becoming an
idea and an aspiration that inspires and connects different
pro-Russian activists, fighters and politicians.

New Russia was an anachronistic vision of new horizons; a
place of the future from deep in the past. But it carried more
than promises; it carried an ethnic and national sense of identity and destiny, powerful tinder which has been reignited in
the early twenty-first century. New Russia does not just reflect
a Russian or a Ukrainian crisis. The fragmenting and discon-
tented territorial politics of our era are creating ever more fluid
political maps. These centrifugal forces are driven by a deep
sense of injustice and anger, and will not be easily appeased.

The Sovereign Military Order of Malta

Walking along the darkly shaded and deeply fashionable Via Condotti in Rome, not far from the Spanish Steps, one might notice something special about number 68. Above its imposing gateway hang two flags: they both have red backgrounds, and on one is a Maltese eight-pointed white cross and on the other a simple white cross. On the door is a burnished entrance plate which reads: 'Sovrano militare ordine ospedaliero di San Giovanni di Gerusalemme di Rodi e di Malta'.

You are standing in front of the smallest country in the world, the Magisterial Palace of the Sovereign Military Hospitaller Order of Saint John of Jerusalem of Rhodes and of Malta. It is a mere 64,583 square feet, but has diplomatic relations with 106 countries and maintains dozens of embassies around the world. The US, China and India don't recognise it, but nearly all of the nations of South America, Europe and much of Africa

do. The order has a permanent observer mission to the United Nations. Permanent observers have access to most meetings and official documentation. It also has a representative in place at numerous other international organisations, from the Inter-American Development Bank to the World Health Organisation.

If statehood is measured by international recognition, then the Order of Malta is clearly a state. It even has an army, which operates as a distinct medical unit within the Italian army, and runs its own hospital train with 192 beds. The order no longer has an air force, but it once did. In 1947 Italy transferred many of its warplanes to the order so as to get round a post-war ban. Today all that remains of the order's air force is a single bomber, emblazoned with the Maltese Cross, on display at the Italian Air Force Museum.

Determining the size of the Order of Malta has never been straightforward. From the year of its foundation in 1048 in Jerusalem, it has had many important but scattered territorial possessions, including Rhodes (between 1310 and 1523) and Malta (between 1530 and 1798), but also numerous other towns and castles, as well as four of the smaller Caribbean islands (Saint Christopher, Saint Martin, Saint Barthélemy, and Saint Croix). All these territories are long gone, and in 1834 the order moved to its present and sovereign headquarters. It has another address in Rome, the Magisterial Villa on the Aventine Hill, the home of its Italian embassy. Today the Order of Malta has more than 13,500 Knights, Dames and Chaplains, 80,000 permanent volunteers and 25,000 employees, most of them medical personnel.

But is it really a country of 64,583 square feet? It turns out that trying to answer that question isn't easy. Indeed, trying to work out the physical size of the Order of Malta, as if it were just like any other state, may be a mistake. It is unique. It does not claim to have any territory at all. It does, however, claim to be a sovereign state and a 'sovereign subject of international law'. To put it another way, the order proclaims that it is an independent entity and it is widely recognised as such. But it does not claim to be a country. If any of these distinctions are clear to you then you are doing better than I am. They throw up more questions than answers. After all, what is sovereignty without place, without somewhere to be sovereign over? When exiting their front door on the Via Condotti, to what government do the denizens of the order owe their allegiance?

When I first came across number 68 on the Via Condotti, I was wandering from one tourist trap to another. I was especially pleased when I saw those flags, and assumed that I'd come across a delightful anachronism; an ancient institution out of place in the modern world. But now I'm not so sure what I found. There is a deeply challenging and novel geopolitical idea at work at number 68. The notion that sovereignty can be based within a network of people and not defined by borders isn't easy to get one's head around. It certainly excites specialists in international law. The Order of Malta is regularly cited as a legal oddity. A 1991 judgement from an Italian court ruled that it was 'the holder of a system of its own' which possesses a 'peculiar international subjectivity'.

This curious formula could also, perhaps, once have been applied to the Holy See, the ecclesiastical jurisdiction of the

Catholic Church; though the Holy See has since acquired more conventional national form in the shape of Vatican City, created in 1929 as an independent city state.

Some have argued that international institutions, like the European Union, have sovereignty without nationhood. And that the same can be said of countries under foreign occupation who have a widely recognised government in exile. But I'm not convinced by these comparisons. They are more like supplements to conventional territorial claims than what we find on the Via Condotti. Maybe a more convincing comparison is with virtual micronations; dreamt-up kingdoms that exist as reams of text on the internet. But these are as ephemeral as soap bubbles, popping once their bedroom-bound creators get bored. And they do nothing. The Order of Malta is over a thousand years old and is recognised internationally both formally and for the medical, social and humanitarian projects it runs in 120 countries.

The Order of Malta is a geopolitical head-scratcher. Even more so since its odd status seems to help rather than hinder its activities. Its stated aim is the 'defence of the (Catholic) faith and assistance to the poor'. Despite its overt religious intent, it has the kind of global access rarely granted to more conventional states. As well as its worldwide relief missions, the order runs the premier maternity hospital in the Bethlehem region, a field hospital in Kilis on the Turkish border that takes care of Syrian refugees and a children's hospital in Aleppo, Syria. In 2016, over 31,000 people fleeing across the Mediterranean were given medical and humanitarian aid by the order's Italian volunteers.

The order's role as a modern agent of faith and charity sits rather oddly with the baroque titles worn by its officers. 'Vested with supreme authorities' over the order is its Prince, Grand Master and Fra' (brother). Until January 2017, the seventy-ninth incumbent was an Englishman, Fra' Matthew Festing. The Grand Master lives in the Magisterial Palace in Rome and is elected for life 'by the Professed Knights by the Council Complete of State'. A period of unprecedented turmoil and papal meddling in the modern leadership of the order abruptly curtailed Fra' Festing's tenure. In 2017 Pope Francis intervened in a dispute between Festing and his third in command, Albrecht von Boeselager, the order's Grand Chancellor. The Pope requested Festing's resignation and Festing's response was, in his words, 'OK, fine.' 'You have to accept discipline,' Festing explains; 'you have to accept obedience. It's the way things are.' Although some have seen this as a Vatican invasion or takeover, Festing's submission to papal will is more a question of adherence to Catholic doctrine than of one head of state surrendering to another. As I write, the order has a temporary, one-year Grand Master, Giacomo dalla Torre del Tempio di Sanguinetto. Over the coming years, it will be interesting to see how much sovereignty the Pope allows the 'sovereign order' in choosing its own leadership.

The Grand Master possesses an Order of Malta passport, as do the Grand Commander and the Grand Chancellor. No other 'ordinary passports' are issued, thus making these three easily the world's rarest internationally recognised passports. A further four hundred 'diplomatic passports' are issued to the order's staff around the world. The bureaucratic paraphernalia of this state-without-a-country is issued from the Magisterial Palace. It is from this address that the official currency of the order is issued. It was established in 1318 and is called the scudo, a coin which is subdivided into smaller denominations, the tarì, grani and piccioli, and pegged to the euro. The coinage is essentially a collector's item, the scudo only being recognised tender within the order itself. However, the order's postage stamps are accepted by nearly sixty countries, from Cuba to Mongolia.

The curios of this curious state are part of its charm. But the Order of Malta is far from being an anachronism: it is a major worldwide provider of charitable and medical aid. Every year it helps hundreds of thousands of people. And it is a serious state, serious enough for conspiracy theorists to ascribe it regularly with enormous power and a secret intent to infiltrate the world's governments in order to take back the Holy Land. This story has been picked up by Islamist jihadists, who have called on their followers to attack the order's embassies. The order's embassies in the Middle East have had to issue statements denying these defamatory rumours.

At one level, the role and nature of the order is easy to understand: it's an ancient Catholic organisation that helps the needy. But there is a mystery here. It's not a mystery concerning the order's secrets but its very public and obvious oddness. Here is a state without a territory, and a sovereignty without borders. This is the smallest country in the world – it does have those 64,583 square feet at Magisterial Palace – but in a variety of ways it is also the largest. It covers just a tiny patch of ground, but its activities reach anywhere and everywhere.

The Stratford Republic

On 23 June 2016 the United Kingdom voted to leave the European Union. On 8 July Stratford Grove, the street I have lived on for twenty-three years, declared its independence from the UK. We occupied one end of the road and set up a stall where, after passing a citizenship test, passports were issued and stamped. I have mine beside me: a crumpled bit of paper with an ink-stamped cockerel motif and the proclamation 'The Stratford Republic requests and requires all those whom it may concern to allow the bearer to pass freely without let or hindrance anywhere they want to go!'

A scrap of paper; a memory of a street party. Nothing more. But it's odd how hard it is to toss it away. There's an alchemical quality about a passport – any passport, even this one. Ink and paper are transmuted by ceremony and turned into something else. It's the ritual that counts, no matter how daft. One of the questions on our citizenship test was, 'Should the Stratford Republic develop a space programme and colonise the stars? Yes or no?' Everyone voted yes. Our national song is 'Boogie Wonderland'. The sun came out, a stiff wind flapped the bunting festooned with felt-tipped flags and lots of people came. There were games, such as 'pin the UK on the EU', where blindfold children attempted to blu-tack a picture of the UK on a map of Europe chalked on Tina's yard wall. And endless border-making on the tarmac. We found ourselves with an ever wider and more elaborate territory. There was also a national flower competition: the eradicable ivy-leaved toad-flax, which has tiny purple flowers and sprouts off every wall, was the clear winner.

Wolfgang and Gaby had brought out big planters to stop the traffic getting through, and for six hours or so we clotted into an excited gaggle, ideas about what we could achieve flying back and forth. People wandering by joined in. I had an earnest conversation with an Italian professor who saw our flags and borders and pinned me by the drinks table, trying to glean what the English were up to and what they thought would happen next. I'm not sure I was much use to him, but I did give him a wine-box-induced hug.

It was a happy day. Made more so for me because I was nervous about how it might end up. I'd been warned of trouble; that our party would be invaded, that fights would break out; and that the whole topic was too contentious to be played with. I'd spent the previous two weeks or so dropping in on my neighbours, trying to drum up ideas and enthusiasm. That wasn't hard. Though, perhaps thankfully, the 'tug of war' between 'Remainers' and 'Leavers' and scream therapy ideas didn't materialise, nor, unfortunately, the garage-door showing of the classic British comedy of urban secessionism *Passport to Pimlico*.

Some of the children in the street are still beavering away at ideas for the new statelet, and I still get postcards and letters addressed to 'The Stratford Republic', although they risk getting redirected. We're not the only street or area in this city to have had the idea. The surrounding district, which is called Heaton, has recently spawned a cottage industry of people producing flags and mugs and t-shirts advertising 'The Heaton Republic'. Brexit seems to have tripwired an interest in urban secessionism, and the realisation that the scale and shape of government can change in the blink of an eye.

It used to be said that people's connections to the small scale were weakening, that local attachments are dying in an ever more uprooted and mobile world. I don't think that is the case. In fact, this argument can be turned on its head: aren't the local places, the ones that are truly ours, the ones that matter most in a disorientated world? It is the bigger political units, the international giants like the EU as well as the national ones, that seem to be most in trouble. They certainly attract most of the blame. The Brexit vote reflected this distrust; the endlessly recycled conviction that there are big, faraway struc-tures of government that are responsible for all the bad stuff that happens. It's the modern mantra: 'they' are to blame and us 'ordinary people', 'the public', are ignored and voiceless. It's a grumbling world view that leads to some curious outcomes; such as the fact that when Brexit won, no one knew what to do

next, or even seemed that pleased. People had voted *against*
something, not *for* anything. There were no street parties in
celebration of the result. Living in the North East of England,
a heartland of the Leave vote, you'd expect that I'd have come
across at least one; some sign of celebration. But there was
nothing: just stunned silence.

The UK has propelled itself to the forefront of our era of
fragmentation. That's very tangible in my home city, in
Newcastle upon Tyne. We are 40 miles from the border with
Scotland, where the vote was solidly Remain. Step back down
across the border into the northernmost part of England, and
the vote was overwhelmingly Leave (with the single exception
of the city of Newcastle). Overnight that border became much
more real. It's a major shift, and it's happening around me.
This book has been following our era's centripetal forces all
over the world, but when they come home, when it's your own
country that is coming apart at the seams, then it's not just a
set of ideas; it's visceral. You start to feel that strange mixture
of liberation and fear, freedom and bewilderment that accom-
panies upheaval; it's frightening, giddy; you worry, then you
just abandon yourself to fate; it sets the heart pulsing, the
blood rushing.

In the months following the vote, the 'Leavers' and
'Remainers' began to be framed as two antagonistic tribes who,
if internet chat is any guide at all, hate each others' guts. That's
the post-facto myth. I'm a good example to show why it is
nonsense. For, even though I organised a pro-EU street party,
I never liked the EU. Its lack of democratic accountability and
culture of homogenisation and centralisation always made me

squirm and, convinced that a 'Leave' vote would be destructive and would never happen, I spent months telling everyone that I wouldn't vote. I got more pro-EU the closer the vote got, voted Remain and was amazed and upset by the result, but I cannot claim much credit with either side. I'm in the happy position of being someone who can be despised by both camps.

But I suspect that this, or something like it, is the kind of queasy, ambivalent position that a lot of people occupy; and if any voice has been ignored in the rush to divvy up the UK population between Leavers and Remainers, it's the voice of us swivelling self-doubters.

Dividing the British up between the Leavers and Remainers, the pro and anti-EU, also ignores the fact that the geographical consequences of Brexit are better characterised as fragmentation and secession rather than 'British independence'. In the wake of the vote, the Scottish First Minister Nicola Sturgeon spoke to London Mayor Sadiq Khan about how they could remain in the EU. That Scotland will eventually leave the UK and make Brexit an English exit or, more precisely, in this complicated union of territories, a RoBrexit (Rest of Britain exit), appears more likely than not. Elsewhere the claims to secession are less robust but they are getting a good airing. There is a petition with, the last time I looked, 179,843 supporters, calling on the London Mayor to declare London independent from the UK. The idea is that London could be a city-state, like Singapore. Leading city businessman Kevin Doran said that London becoming an independent state is not just possible, but inevitable 'within twenty to thirty years' time'. Peter John, the leader of the

London Borough of Southwark, where 72 per cent voted in favour of remaining in the EU, said it was time to look seriously at London's political relationship with England. 'London would be the fifteenth-largest EU state,' he pointed out, 'bigger than Austria, Denmark and Ireland, and our values are in line with Europe – outward-looking, confident of our place in the world.'

This kind of talk does nothing to endear London to the rest of England. Fresh wedges are being driven into the body of the United Kingdom. The eminent sociologist Richard Sennett even claims that 'London belongs to a country composed of itself and New York.' Some in the capital will nod sagely. But it's a bizarre remark, in keeping with the kind of separatism that has become commonplace among 'Londonistas'.

There is a febrile and dissatisfied quality to the public mood, which is energising but also unpredictable. Perhaps this is what most of the world has been living with for generations; an endless sense of imminent destabilisation. I can't claim much knowledge of the world: I've travelled more than most, but my travels have been superficial and transitory. What I do know is this street where I've been going to bed and waking up for so many years. It's just a set of shut doors and small gardens and yards, but it's where I think of when people talk about community. The daily conversations, the smiles and waves; every once in a while the shared cup of tea. This is what I know and know deeply. The world fragments; the kaleidoscope shifts; but I keep plodding down to the corner shop, same as usual. There's Tina; it will be her birthday party soon. There's Jack, parking his van. This is my small, familiar patch. A few months ago, we had a party; we will again. On the larger scale of things it didn't matter; but it mattered to me.

III.
Utopian Places

Utopia has a close yet uneasy relationship with place.
Thomas More's *Utopia* was an ideal place, but it was
also a non-place. More's island of Utopia was a fable, even a
kind of mockery; somewhere that could never be. His vision
looks rather sinister today; rule-bound and serviced by slaves,
it forces us to realise that the perfected society might not be
perfect for all. Utopia has a dark side: Nazi Germany was a
utopian experiment just as much as any hippy commune. We
begin our journey with the bleakest possible ideal society –
the Islamic State of Iraq and the Levant – before heading off
to the happier terrain of freedom-loving eccentrics. Utopia's
many faces all gaze lovingly at place: they pour their efforts
into shaping and imagining the best of all places. For denizens of Cybertopia, place has been virtualised into an online

realm; but it is still almost quaintly traditional in its insistence
on creating recognisable landscapes; often with bucolic paths,
islands and towns. The wandering, pod-dwelling New Nomads
are another group of new utopians. Their weightless vision
of the world rests on the idea that happiness can be found in
mobility. I've never found this an appetising prospect; turning
life into one long road trip might work for the New Nomads,
but it's not for me. It was precisely attention to one small place,
an expression of loving care, which distinguished Nek Chand's
labours in the Indian city of Chandigarh. I'm fascinated by the
contrast between his Rock Garden, a secret world of waterfalls and bowers, and the functionalist, modernist city that
engulfs it. Leaving India, we travel through two more creative
and quirky utopian urban enclaves, Christiania and Helsinki's
urban foraging community, before returning to starker and less
comfortable visions of the future. The chapter entitled 'City of
Helicopters' takes us to São Paulo's skies, and 'City without
Ground' wanders through Hong Kong's elevated living and
working spaces. In both cities a new vertical and three-dimensional urban reality is being introduced, leaving the earth far
behind; at least, for those who can afford it.

The Islamic State of Iraq and the Levant

Two small boys are standing on an American flag, pointing up at a gigantic map of the world. On the map great splashes of bright light are erupting from North and West Africa and across the Middle East. It's a picture from the cover of a school geography textbook issued in the Islamic State of Iraq and the Levant (ISIL: the Levant refers to the countries of the Eastern Mediterranean; ISIL is also known as the Islamic State of Iraq and Syria, and Daesh), and the boys are looking at parts of the world ISIL controls. If this textbook is any kind of guide, then the confidence and global ambitions of this ferocious new nation appear to know no bounds.

The map of ISIL was always unclear. We are used to defined borders, and plenty of maps of the region imply they exist, but in fact all ISIL's borders are dotted lines, uncertain and shifting. As I'm writing this, in June 2017, ISIL's map is being torn to

shreds; its Syrian capital of Raqqa is about to fall and it controls
only a few streets in what was once its greatest conquest, Iraq's
second city: Mosul. The Islamic State are despised by most
people in the 'Arab World', just as they are in the West, perhaps
even more so since their brand of Islam is directly blamed for
economic collapse and for staining the reputation of the faith.
Yet in military terms, ISIL's achievement has been considerable. At its height, in 2015, ISIL was about the size of Jordan.
It had a population of 10 million and stretched across the
Syria–Iran border, connecting its de facto capital of Raqqa in
Syria to Mosul and down to farming towns south of Baghdad.
It lost territory throughout 2016, and its area of direct control
was reduced to thin and disconnected strips.

But still it lives, for the time being at any rate, so I'll be using
the present tense. ISIL has a good claim on being the most
ideologically extreme example of state-making the world has
ever known. It is also the most extreme illustration of how the
desire to create a pure, perfect and reborn society – a utopia
– can give birth to monsters, and how place-making can mix
both nostalgia and genocide. This is a theocratic state with
violently totalitarian ambitions: it is defined by acts of mass
sacrifice, mass slaughter and mass destruction.

Many find some comfort in the idea that because it is so
utterly merciless and malevolent, ISIL is not really a religious
state; that its followers are only pretending to be faithful.
But this isn't true. Almost all its actions, and every law it
passes, reflect the depth of its religiosity. The head of security
for the Kurdistan Regional Government in Northern Iraq,
Masrour Barzani, recalls how one would-be suicide bomber,

arrested moments before his attack could succeed, shouted out in genuine grief: 'I was just ten minutes away from being united with the Prophet Muhammad!' For Barzani, 'They think they're winners regardless of whether they kill you or they get killed.'

Founded in Iraq in 1999 by a Jordanian ex-criminal, Abu Musab al-Zarqawi, ISIL adopted its current name in April 2013 after expanding into Syria. Expansion is central to its ideology and state practice. It wants to replicate itself across the globe, and this desire is an urgent one. Even when it is losing ground ISIL is mapping out its next phase. It believes that, with the help of God, it will expand to include every land that has ever been Islamic. This will involve the retaking of Spain and Portugal, drawing both countries into a caliphate that will encompass the whole of West, North and East Africa and stretch through Turkey and Saudi Arabia, the Gulf States, Iran, Pakistan and on into China. And after that? The conquest of the world.

ISIL is not formally recognised by any nation, and rejects all international and national treaties and organisations as infidel blasphemies. It divides the planet into *Dawlat al-Islam*, the State of Islam, and *Dawlat al-Kufr*, the State of Unbelief. The latter can expect no mercy. One of the reasons it does such terrible things, like blowing up a Russian aeroplane full of tourists and carrying out shootings and stabbings of ordinary people in Paris and London, is that it hopes, and prays, that infidel armies will retaliate. For, according to the prophecy its leaders cling to, a final clash of believers and non-believers will hasten Judgement Day. ISIL is an apocalyptic project, propelled by a firm faith in a fast-approaching End of Days and a complete absence of doubt, introspection or even simple curiosity.

Its bullheaded certainty makes it a ferocious opponent, but also sows the seeds of its vulnerability. For ISIL is not only hated, it is also a sitting target. The roving and footloose activities of earlier Islamist groups, like Al-Qaeda, made them hard to attack. Having a fixed territory, with all the infrastructure of a normal state, with roads, electricity stations, factories and geography lessons, makes ISIL an easy place to disrupt, especially from the air. Yet it carries on, in part because of the constant stream of new recruits it attracts. ISIL has become a talismanic destination; somewhere true believers from across the world journey to, convinced that it is their duty to live in, and die for, an authentically Islamic society. Security intelligence experts The Soufan Group estimated in 2015 that between 27,000 and 31,000 people from 86 countries have travelled to Syria and Iraq to join the Islamic State and other religious extremist organisations.

ISIL has to recruit, build and maintain just as fast as it blows up, crucifies and beheads. By 2015 it had developed a system of regional government, with governors in Iraq and in Syria, with each region having its own officers in charge of prisons, security and economic targets. Islamic advisers and religious specialists operated or had a hand in nearly everything that went on, and the whole structure was directed by ISIL's all-powerful cleric, 'Caliph' Abu Bakr Al-Baghdadi. It was a top-heavy operation (which is unlikely to have survived the defeats of 2016 and 2017), entirely geared around drawing resources and manpower into military use. However,

ISIL also had an extensive if primitive welfare system, with free healthcare and bundles of cash handed out to the 'deserving' poor.

The benefits of living under such a regime are highly insecure. The lines on the map move fast: enclaves and territories can come and go overnight. Combined with the ferocity of the warring parties, these fluid borders mean that among ordinary people a sense of uncertainty has taken root about which direction is safe and who they should trust. The region's hydra-headed civil wars have produced a cartographic patchwork that resembles medieval maps from a time when landownership was broken up into myriad kingdoms, local loyalties and clans. However, this modern patchwork does not reflect the influence of pluralism so much as its eradication.

Lebanon, Syria and Northern Iraq were once the most religiously diverse places in the Middle East, with ancient communities of Christians and Yazidis living alongside different Islamic minorities. The importance of place to these groups has been made painfully clear as they have been driven into safe havens and forced into exile. One of the most pitiful examples was seen in August 2014, when Yazidis fled from ISIL forces who had branded them 'devil worshippers'. About forty thousand climbed for safety up Mount Sinjar in northeast Iraq, a site holy to the Yazidis since it is where they believe Noah's Ark came to rest. After weeks stranded out in the open, most eventually escaped, hoping to join the many thousands who, like the Christians, Jews and other minorities before them, had fled to Kurdistan, Turkey, Europe or North America.

ISIL wants to create a pure place, an Islamic utopia free of
complexity. Against the tide of global multiculturalism, it offers
a return to the certainties of monoculturalism. The bulldozing
and sledgehammering of ancient sites, shrines, Shia mosques
and churches is a response to the threat that ISIL leaders see
in diversity. It is a sad reminder that religious extremists have
an extreme relationship to place; that they like places to be
cleansed and compliant, with nothing standing that questions
the completeness of their territorial seizure and no one around
who does not fit their narrow model of the ideal citizen.

ISIL is so destructive, of itself and everyone else, that it is no
surprise people in the region and the wider world can't quite
believe it exists. The Middle East is awash with conspiracy
theories about its origins and financing. The Lebanese foreign
ministry took these rumours so seriously that it summoned
the US ambassador to respond to the accusation that ISIL
was a US and Israeli plot. Picking a newspaper from the news
stands in Cairo, I read that ISIS 'are acting as America's Foreign
Legion', and that the Obama administration 'supplied ISIS with
weapons'. People prefer the idea that Israel and the Americans
are ISIL's puppetmasters to the alternative. Bewildered and
powerless people get an illusion of power from conspiracy
theories. They imagine they have secret knowledge when,
in fact, they have nothing. The first step towards healing the
region's shredded map is to acknowledge that ISIL's puritanical
mission is not a foreign import. It is a mutant, savage child of
the land it is ripping into pieces.

Cybertopia

Twenty-first-century dreams are pouring into Cybertopia. It is a vast, heavy flow, an endless cataract that lifts all along with it and fills our screens with endless chatter about where we are heading, about the delights and the monsters round the next corner.

On the screen window behind this one I'm initialising myself onto Second Life, the largest of many interactive multi-user platforms, and there I am, on the green grass, an immobile humanoid standing beneath someone else's avatar, a human butterfly with sparkling black wings. It hovers annoyingly close while conversational openers scroll down the screen. 'You do know you're naked?' The site promises 'a world with infinite possibilities' and 'a life without boundaries', but I appear to be underprepared. It's my first time on Second Life, and after urgent tapping I acquire some leggings from a shop – for a nominal fee – and a shopping bag appears in my hand. Perhaps I can now spend some time with James2003love ('want to have sex') or Mockingay223 ('sometimes I wish everything was just a dream') and all my other new friends. But the shopping bag won't open, so my grey-brown body has to roam this world like something damned, fleeing one purgatorial circle after another.

I fly through the blue sky and trip lightly down on Disco Island. Do I want to dance? I suppose so. How was I to know the fiend who rules this land has turned this optional click into a permanent state of being? In an instant I'm bound in adamantine chains of disco and have to foot-slide, hustle, strut and chicken-wing my way into the sea just to get a bit of peace of mind.

So there I am, beneath the waves with my shopping bag, naked, flailing and feeling like I'm missing out. Because the geography of Second Life is intoxicating: across these endless horizons identity and landscape go hand in hand. The fantasies of Cybertopia take time to unfold and have an imaginative hold because they take the issue of place seriously. Scrolling down Instagram posts, checking out Facebook, is fast and on the go because these sites have no 'where'; users' investment in them is non-immersive. It is only when you slot in trees, islands, buildings, and allow people to wander around and through them, that the cybersphere begins to feel sort of real.

Place-making does, however, come with a price tag. Land rents on Second Life are so expensive that a bunch of cheaper versions of the same idea are now on the market. Having your own land is the ultimate prize. In the top-right corner of my screen I'm being pestered to acquire 705,420 square feet, which costs 'only L$1,499 per week!' (that's US$6 in real money). This would buy me "'The Cove" Hyper-Realistic Sky Dome by Bella Pointe! Total privacy, No visible neighbors!' It does sound like a bargain, as most islands cost over US$400 (in real money), and the site's owners boast that 'well-off individuals' have spent US$10,000. The 'Massively Multiplayer Online' game *Entropia Universe* attracts an even wealthier crowd: back in 2009 Buzz Erik Lightyear spent US$330,000 (again in real money) for Crystal Palace Space Station, and more recently a Los Angeles-based games company paid US$6 million to obtain their very own planet.

Why join a virtual interactive community and then buy a private domain, one where you don't have to be bothered by the 'visible neighbors'? It sounds like a good question, but at the moment I'm the one hiding beneath the waves. I know they are waves because I can hear their faint roll. Yet the loops of unobtrusive sound effect can't disguise the fact that Second Life, along with all the other 'virtual reality' sites that jostle for our attention, feels about as real as being in a Tex Avery cartoon. Crudely jointed puppet people plod across colour fill-in and fly over cut-and-paste flatscapes of grass, sea and stone. The participants' tentative and tiny stubs of text resemble not so much acts of communication as incantations, little spells trying again and again to give these dead pixels some life. Arriving later at the Islamic holy place of Moat Mount, which is blanketed in birdsong, I find everyone sitting in a circle moaning about how 'laggy' they feel and how 'laggy the land was'. I thought they were sitting still in meditation but, in fact, it was because the website programme was misbehaving, and every time they moved there was a jolting mismatch between movement and landscape, leaving limbs and buildings hanging in mid-air. After a minute or so, an unshaven man in a beanie, with a cigarette poking out from somewhere near his mouth, pushes me into the sea (something you do by bumping into people). It seems to be the only place I'm fit for.

Maybe if I had paid for an Oculus Rift headset, or just waited for the new version of Second Life to come out, that might all change. But even the youngest of us is old enough to know that the totally immersive, *really* real experience is always just round the corner. It has never arrived, and it looks like it never will. And the prospect of virtual reality feeling more real is already generating a backlash among Second Lifers: many complain that they enjoy the site precisely because it doesn't look or feel too real, and point out that if 100 per cent reality was what they were after they could just open the door.

Richard Barbrook of the University of Westminster argues that we have witnessed not the evolution but the *repetition* of cybersphere's utopian promises from the 1950s onward. In *Imaginary Futures* he details how the 'hi-tech utopia is always just around the corner, but we never get there'. In other words, the rhetoric of a super-advanced, interconnected digital world has become a stuck record. 'The future is what it used to be', says Barbrook, and it's a vision with an ideological agenda. That agenda is rooted in Cold War attempts to identify capitalism with liberty; of endlessly churning out the myth that technology will soon be liberating us all and offering untold excitement. For Barbrook this is the 'strangest story' of our times, made weirder by the way so many of us are ready to believe these recycled promises.

Perhaps, though, we can go back further than the 1950s, all the way back to the first virtual world, the place that all subsequent ones obliquely refer to: the Garden of Eden. Cyber consultant Margaret Wertheim, in *The Pearly Gates of Cyberspace*, sets out this link to Eden and the transcendental hopes that are projected into the cybersphere. Calling cyberspace 'a repository for immense spiritual yearning', she argues that it provides a secular pathway to heaven; 'a yearning for transcendence over the limitations of the body . . . a longing for the annihilation of pain, restriction, and even death'. These are hopes we see in many varieties of utopianism, but Wertheim's study of how people willingly lose themselves in the cybersphere, giving up their minds and bodies to it, presents a convincing portrait of how it offers a new kind of spirituality, complete with salvation and the promise of 'a place *outside* space and time, a place where the body can somehow be reconstituted in all its glory'.

With spiritual glory in mind I have, after much effort, managed to extract myself from the waves and flown over to the Living Memories Memorial Garden, a Second Life sanctum where people can 'claim a tree with a candle and plaque' in memory of the dead (real people who are really dead). But I'm saying goodbye in another way: Cybertopia may be the utopian project of our time, and it may even offer some kind of transcendence, but I'm not attracted by its promises of virtual life. The novelty of the virtual wore off years ago and, heading towards the 2020s, it's beginning to feel like nostalgia for the future. And it leaves me, all too frequently, frustrated; as now, because my undressable avatar, handbag still flapping, has got trapped in forward mode. This often happens, but somehow this is worse, as I trample and retrample a memorial tree in full flower. I would be doing so still if I hadn't closed the programme, unregistered my avatar and escaped back to reality, wherever that might be.

The New Nomads

Place matters, and having a place called home is a basic need. This may sound obvious, but increasingly it needs to be said because there is a powerful new utopia that challenges it. This vision claims that the untethered and ultra-mobile life is the good life. It's an alluring dream that meshes global capitalism's culture of footloose, 'hire and fire' flexibility with progressive values, like cultural openness and hostility to dull routine. I'm more in thrall to this shining destination than I care to admit, which is why I bought a beautiful book that extols its virtues: a book of photographs of handsome people sitting in unusual tents and tiny, wheeled huts. It's called *The New Nomads* and I got rather obsessed with it; a romance that ended when I made another aspirational purchase, of a tent that hovers above the ground. But I'll get to that fiasco later.

The front cover of *The New Nomads* shows us a shallow lake at sunset in which is parked a sleek, pod-like caravan, the water rippling at its hubcaps as well as around the shins of a young couple, the proud owners, standing hand in hand. When I first glanced at the cover I assumed this was a worthy volume about flooding and climate change; that these were stoic survivors. But no; this is a photograph of the 'Sealander', one of the bespoke mobile homes that are being lovingly crafted for a fresh-faced tribe of well-heeled go-getters.

> Mobility is the new ultimate form of freedom . . . Today's creative class thrives off a lifestyle that enables it to work six months in a shared office in Berlin, spend the summer in a caravan in Chile, and show up just in time for the next project at a temporary desk in New York.
>
> (*The New Nomads*, back cover)

The New Nomads are young, entrepreneurial globetrotters who are at the centre of a growing market for pared-down and portable living spaces. These range in size from gorgeous bare wood cabins that can be collapsed and transported away, down to the 'Sack und Pack', a 'minimalist product for the contemporary wanderer' that consists of an oak stick, a strip of leather and a square of cloth. 'Holding just the most elementary belongings', we learn that the Sack und Pack 'frees us up to embark on an authentic and adventurous journey'.

The elegant youth shown posing with his 'Sack und Pack' is a treat of hipster fashion: cardigan, flat cap, long garter socks and half-length trousers. I'm not sure why, but I find the collision of high style, high-mindedness and tiny spaces incredibly funny. I particularly like one photograph in *The New Nomads* that shows a young woman inside her pod home, which resembles a large and flimsy white drum with paddle attachments. She has wedged herself into this tiny object, somehow managed to turn round and in a foetal position stares out, Zen-like, with a sort of 'I am complete' expression on her face. Perhaps she uses it to splash to and from her 'temporary desk' in New York or her 'shared office' in Berlin.

I also can't help a guilty, place-bound smile when reading about the 'digital nomad' Jan Chipchase. 'Author, adventurer, product designer, and global entrepreneur' Jan 'negotiates upwards of 40 cities a year, answering pressing branding and research questions from his Fortune 500 clientele'. He is also the inventor of the 'Raw Utility Pouch', specially made for 'transporting US letter-sized documents through treacherous conditions' and the D3 Duffel, squashy aeroplane hand luggage inspired by Chipchase's personal '"no wheels" policy'.

It is remarkable that the New Nomads have managed to establish such a fashionable image from what is, ostensibly, such unpromising material. Strip away all the designer verbiage and what we are left with are caravans and bags. There are certain words that are banned in this new realm, words that dare not be spoken: 'trailer park' and 'caravan site'. Mobile home parks are some of the oddest places on earth. Both permanent and vacation parks aspire to a lightness of touch; they hover above the ground, just waiting to take off. Yet they couldn't be less 'New Nomad'. Those ranks of semi-residential boxes; the clipped pathways; the little flags; usually some form of clubhouse: they look constrained and tethered, a clotting-together that suggests community is being prized as highly as mobility.

It's a comparison that highlights the ultra-individualism of the New Nomads. Here uniformity is anathema, and where congregations do occur they are small, spontaneous gatherings at a festival or around an open fire somewhere far away from the crowd.

I don't doubt that the New Nomads are as happy, fulfilled and creative as *The New Nomads* keeps telling me they are. They do, however, make some irksome assumptions. The explicit message found throughout the 'New Nomad' subculture is that not only do they know how to live well but that the rest of us don't; that we are dull, uncreative folk and dumbly uncomprehending as they zip by in their variously sized pods. There seems to be a lack of insight at work here; or of simple empathy. There are good reasons why people have not wanted to spend their lives rushing from pillar to post; why they have sacrificed and struggled for the very opposite, for a stable, caring community and for a place called home. These affluent nomads cast themselves as creatives and change-makers, but they live in a bubble in more ways than one.

Even more exasperating is the misuse of the word 'nomad'. Traditional nomads do not zigzag rootlessly across the planet or disparage home and place. They move between particular and established places – between summer and winter grazing, for example – each of which is understood intimately. Real nomads have far more in common with place-bound, sedentary people than they do with jet-setting digital entrepreneurs. Maybe if the 'New Nomads' called themselves 'small business executives with tiny caravans' it would be nearer the truth, though I can appreciate why they don't.

I'm kind of jealous, of course. 'Designing a new banking service in Myanmar today, and launching a new brand in Saudi Arabia tomorrow' (that's Jan Chipchase again); stepping out of a floating egg onto a beautiful shore; sporting a jaunty 'Sack und Pack'; I want a taste of what they have. Enthralled and, admittedly, drunk, I thought I'd found the answer late one night while shopping on the internet. Only £800 for a triangular tree tent. You take a long cord from each corner and ratchet it to a different tree, and that's it: you have a hip den high off the ground. It sounded amazing. The next morning my partner was horrified, literally lost for words: she seemed to think that that was a lot of money for something no one in their right mind would ever use. I was prepared for this: resolutely breezy, I painted a picture of our new, weightless life, casually strapping ourselves to trees – an enviable, roving life spent gazing with Buddha-like wisdom and indifference on the world below.

Things have not worked out in the way I described to Rachel that morning, much to her quiet satisfaction. You need to find trees positioned just so to make the whole thing work. Then, after at least an hour of stretching and clipping, it hangs there looking like an accident waiting to happen. Not that I was in danger of falling far; when I clambered up through the flap at its base, it sagged down like an elephant's nappy, the fat dome of me scraping the ground below. My hover tent went back in its bright green carry case and has been lying lead-like in the upstairs eaves ever since, a reminder of my lack of perseverance, the dangers of combining red wine with midnight shopping, and the time I pretended to myself and to the wider world that I was a New Nomad.

Nek Chand's Rock Garden

This is a tale of two cities. In the 1950s two men, one an internationally revered architect, the other a young local road inspector, built two utopias. Le Corbusier and Nek Chand laboured cheek by jowl, some 150 miles north of New Delhi, on the baking plains of Punjab. They never met, and the end results look like they come from different planets. Chandigarh, planned by Le Corbusier, is a stark, clean-lined statement of high modernism, a bold and geometrical new regional capital. Nek Chand's 'Rock Garden' was built in the shadow and out of the detritus of that grand project. It was assembled from fragments of tiles, pipes and other bits and pieces. Today this crazy paving ranges over 24 acres of looping walkways, tunnels and waterfalls; all under the blank gaze of platoons of pebble- and mosaic-covered sculptures: monkeys, donkeys, leaping dancers, people with giant hats, cyclists.

Le Corbusier was feted by Jawaharlal Nehru, India's first prime minister, who declared that the new town would be 'symbolic of the freedom of India, unfettered by the traditions of the past'. Meanwhile Chand worked secretly and illegally; his endeavours were only discovered by the authorities in 1975. Thankfully they recognised its importance and gave Chand a salary, the title of 'Sub-Divisional Engineer, Rock Garden' and a workforce of fifty. Far from being 'unfettered' by the past, his Rock Garden is soaked in nostalgia. And although it's now open to all, with a steady flow of both Indian and international tourists, it is so densely mazed that it remains a place of baroque complexity and concealed nooks.

The relationship between these two places reads like a geographical parable. The austere mother and the crazed child; the place of expertise and rationality, where addresses look like a computer code (the city is divided into Sectors, Blocks and Phases), and the organic, dripping fantasy that is cradled between Block C and Sector 4a. For most of the day Chandigarh looks like a de Chirico painting that has been invaded by traffic. Coughing in the exhaust fumes of one of its relentless, sun-baked boulevards, it's hard to resist the promise of the Rock Garden, with its curved lines and happy oddness.

Hidden behind high walls, the Rock Garden turns its back on its harsh parent. I wasn't sure how to get in. An unsmiling guard waves me towards what appears to be a blank rampart topped by bulky concrete birds. As I get closer I find that a number of deep but tiny square holes have been poked through the walls. Behind each, a melancholic ticket seller is silently wedged. The Rock Garden resists ease of access or orientation. Pushing through the iron turnstile, you find yourself in a courtyard dotted with sculptures and a small pool. With a dozen or so other visitors, I mill around thinking how much smaller Chand's creation is than I expected. Furtive glances pass between us; an uncertain disappointment begins to grow. When a couple of women disappear down narrow and half-hidden steps and don't return, there is a polite surge towards that corner. We spy their orange and red saris up ahead and follow, crunching along a dark passage. Suddenly the day is changed; utterly transformed. We emerge through arches to stand wondering at an enormous landscape that looks as if it has been spun rather than built; paths and levels weave at great height and in many directions; a broad waterfall slips down past spirals, crenulations and curls of decorated mortar. To have a place like this suddenly revealed, a place so hidden, awakens an urgent desire to run and play. There is the immediate and slightly giddy knowledge that you can skip off in any direction and find more and yet more; that there are no prescribed routes, interpretation boards, earnest lectures; that this is a place of sensation.

There is a kind of centre to the web, constructed in its later phases, where family-sized swings hang from the arches of a snaking arcade. In his last years Nek Chand would sit on the little balustrade that runs along the top of these arches, at the heart of his kingdom, delighted giggles welling up below him. He died at ninety, in June 2015, and from the adulatory tributes in the press, and the eulogy about his 'fabulous creation' from the Indian prime minister, Narendra Modi, you'd have thought the whole country was in thrall to his quirky personal vision. Conversely, the critical tide turned some time ago against Le Corbusier's Chandigarh. The idea that India ever needed Europeans to show them the future has come to rankle.

Yet we need to look more closely at these places, and at the men who made them. The kind of portrayal that pits the much-loved local eccentric against the authoritarian outside expert has become something of cliché. It recurs across the world. Le Palais Idéal in Hauterives, south-eastern France, is another example. It resembles a tiny version of Chand's Rock Garden, and was built by a postman called Ferdinand Cheval. It was adored by the surrealists and, later, the situationists, who despised experts in general and Le Corbusier in particular. But what would 'outsider art' mean if it were not for 'insider art'; the irrational if not for the rational? Walking out into the warm night from my hotel in Sector 17, into the brightly lit streets crowded with smartly dressed and good-humoured shoppers, I worry away at the idea that 'eccentric places' only delight us because they are framed by more serious ones. Chandigarh isn't a failure: per capita, it's the wealthiest city in the country, and it takes first place in a raft of other indices of social development. People are proud of their town. Its spatial logic has even seeped into the local idiom: the euphemism for death in Chandigarh is 'going to Sector 25', the area that houses the city's cremation ground. Perhaps it is precisely because it is so wealthy and content that Chandigarh desires and allows the Rock Garden in its midst. They may be opposites, but that is why they need each other.

The connections between the two places run deep. Each, in a different way, addresses a shared trauma. Chand was born 50 miles north of Lahore, which is now in Pakistan, but which was once the capital of Punjab. His Hindu family had to flee south across the new border. Chandigarh was needed because partition robbed the state of its capital; it is an attempt to create a sense of pride and purpose among a disorientated and uprooted population. In a much more intimate way, the Rock Garden does the same thing. It was first inspired by stories Chand's mother had told him of exotic lands inhabited by gods and goddesses. This beautiful realm is recreated in the Rock Garden; it is an evocation not just of a mythical kingdom but of the lost land of his childhood.

Chandigarh and the Rock Garden are both attempts to reclaim territory imaginatively; the one looks to the future, the other to the past, but they are equally bold and, for the time being at least, seem to be happy with one another. A pact has been arrived at, one other places might well learn from: that modernism gets dull without a bit of madness; that even the most rational of cities need their zones of escape. It's a lesson that is especially pertinent for India at the moment. For all the warm words that are said about Nek Chand are being drowned out by the din that is modern India, a good part of which is coming from umpteen building sites where endless modernist visions of India's future are being mass produced. In the housing blocks and roads that sprawl over every horizon, it can sometimes look as if what is being thrown up are chaotic versions of Chandigarh. But unlike Chandigarh, there is no room, no allowance, for that other, equally important side of our relationship to place: no safe spaces for the intimate and the eccentric. They never met, but I like to think that Le Corbusier, who was anything but a conformist, would have enjoyed the work of his unknown rival. I think he'd have admitted that what he left behind – all those bits of broken tile and pipe – had found a good home; that the rational city needs and loves its wild twin.

Christiania

Christiania is an island of freedom that sprawls across 84 unexpectedly green acres in the heart of Copenhagen. The labyrinthine lanes of this Arcadian tangle of largely self-built homes and workshops spell out an alluring message, repeated in the Constitution of Christiania: 'Anyone is free to do what they want, as long as it doesn't infringe on the freedom of others to do what they want.'

This breakaway statelet of about a thousand residents has its own currency (the Løn coin can be used in about fifty shops), laws, government and values. Christiania is the most fully realised city-based experiment in libertarian and cooperative organisation in the world. It's been going since 1972, but I got to know it over a few hot September days in 2013. I was lucky to have Emmerik, one of the founder members, and my friend Helen Jarvis, as my guides. Helen is an academic authority on Christiania who seems to know every story and every byway within this freedom-loving enclave.

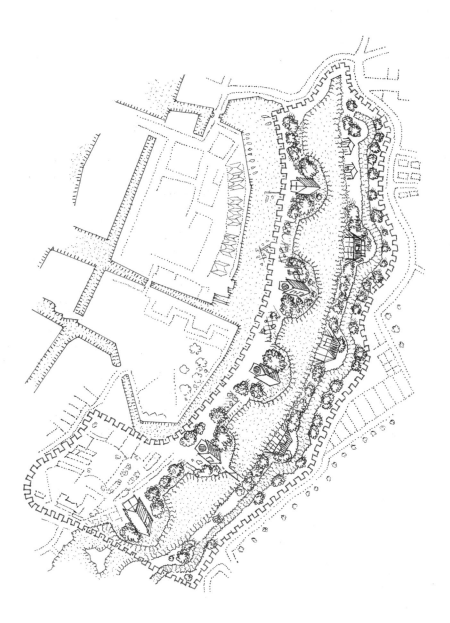

Back in 2013 visitors were met first by Pusher Street, a dope-peddling zone. A few years later, in September 2016, this infamous stretch was marred by deadly violence, shootings that left one man dead and several injured, and today it is gone. It was, in any case, always a distraction from what Christiania is all about; something I learn from Helen and Emmerik as we wend our way to his home in The Dandelion, one of the fourteen autonomous villages that make up Christiania.

Strolling with these two at my side turns Christiania into a snug village; we stop constantly for chats, news and greetings, though much of the foot traffic is made up of curious sightseers. Christiania is Copenhagen's second-biggest tourist attraction (just pipped by the funfair rides at the Tivoli Gardens), and Emmerik readily admits that this pulling power helps explain its survival. Christiania is now a national treasure. We pass a 'leave and take' hut; looking like a rather messy yard sale, it is where residents leave or take clothes and a whole lot of other household items. Like so much else here it is an unstaffed, self-running enterprise. Other services, including water, sewage and general maintenance, as well as the local newspaper and post office, are funded by a modest flat-rate 'use fee' that Christianites pay each month into a collective fund.

Soon we arrive at one of the icons of Christiania, the Buddhist stupa, a white vase-like monument built by a man with a shaggy white beard who is pottering about next door inside a great tent that covers the ark-like boat in which he lives, and which he appears to be permanently building. From here we head down an alley to the 'central store', the Green Hall, a lofty eighteenth-century timber structure that was once a riding arena. It is full of everything you'd need to build a home: planks, nails, doors, windows. It also functions as an alternative supermarket, with a big emphasis on recycled furniture and office equipment. The amount of office stationery disconcerts me: I didn't expect the need for so much pen-pushing in Christiania. But I'm learning that one of the ways individualism finds expression here is through an appetite for small business. The 'hippy' tag that Christianites often play up – along with the notion that they have created a 'losers' paradise' – is misleading. Once you get a bit under the skin of this place, you start to notice its entrepreneurialism, diligence and almost prissy attention to detail. These qualities are allied to a uniquely Danish fondness for 'hygge' (pronounced 'hooga'), the recently fashionable trend word that encapsulates a state of cosy, beautiful homeliness. Copenhagen's shops are stuffed with knitted, carved and woven things catering for the 'hygge' market. Despite its polyglot, anarchistic character, Christiania provides a bohemian version of this typically Danish, comfy ideal.

We pass through an arch into a sunlit circle of wooden houses. Emmerik lives here with his partner and daughter. It's like stepping into the pastoral paradise described by William Morris in his utopian fable *News from Nowhere*. It's a comparison made unavoidable not just by the fact that this enclave of birdsong and flowers exists in the middle of a considerable metropolis, but by the handcrafted, laboured-over nature of almost everything, from the children's swings to the door lintels. Emmerik's walls are crowded with musical instruments and vinyl records, but the house is clearly a labour of love, where the planed surfaces of the tables, windows and banisters attract the touch of your hand. Although Christiania also caters for fully communal living arrangements, and all the land and houses are collectively owned, one of the most compelling ways individualism finds expression here is in these lovingly cared-for homes, places that are unique and feel private, but are also part of a shared endeavour.

Householders seem to pride themselves on the multiple ways they can reuse what the rest of Denmark throws away. It's rare here to see anything new or bought. This approach reaches its apogee in the startling 'glass house', a vertiginous and rather lopsided dwelling that appears to be made entirely from recycled window frames. The Danish architect Merete Ahnfeldt-Mollerup has called Christiania a laboratory for 'adaptive reuse', and it is strangely heartening to learn that its mishmash of self-created structures has brought forth shelves of commentary from the architectural profession, many of which regard it as one of the most exciting built environments in Europe. Other famous Christianite architectural curiosities include the Bananhuset (Banana House), the Pagoda and a tiny house resembling a recently landed UFO that is nestled among the reeds at the water's edge of a large lake.

Emmerik explains that the Danish navy abandoned this site long ago, and in 1970 a section of fence was knocked down by local parents so that they could commandeer a bit of land to create a playground. The same year an alternative newspaper called *Hovedbladet* ('Head Magazine') suggested the reuse of the old barracks, and an influx of squatters began. The existing 150 buildings were soon being added to with about 180 more self-built homes. By 1975, the population had reached about 900, and, unlike the rest of the city, it has barely grown since.

Emmerik, Helen and I end up talking about the controversies that cloud Christiania's sky. Well before the 2016 shootings, this was a divisive place. Even the tolerant Danes, who are one of the few nations open-minded enough to allow Christiania to exist at all, find their patience tested. After all, this happily rule-free zone occupies dozens of precious acres right in the heart of a crowded city. In Christiania there are people who live in quite large, detached houses surrounded by trees and gardens, in a way that only the super-rich in the surrounding borough of Christianshavn or the rest of the city could afford. Christiania's low-density living occupies space that could otherwise be a public park or housing. It is a gated community which operates strict immigration controls: despite having fewer than a thousand residents Christiania is, apparently, full. One satirical Danish TV show *Den halve sandhed* ('The Half Truth') poked fun at this idea by carting in a load of planking to a pleasant lakeside spot and trying to build a small house. It took only a matter of minutes before locals were objecting, angrily dumping the imported wood back in the film crew's trailer. The new arrivals countered that, surely, they had just the same right to build here as the existing residents. But the Christianites didn't agree. It is only with the full agreement of one of the autonomous 'villages' that anyone can move in, and since the locals are happy with their wide acres and semi-pastoral existence this is very rare, happening mostly on a 'one in, one out' basis.

It's not an unreasonable position. Christiania would quickly collapse if it threw open its doors: people would swarm in and the wrong sort of anarchy would reign. The existing residents have built something unique, and they want to keep it going. Perhaps freedom must be selfish in order to survive: the unstated message is that in order for us to be free, you are not free to join us. This hard-nosed aspect of life here is rarely admitted to, which does make it a bit galling when Christianites loudly proclaim the rhetoric of 'no borders, no nations' and position themselves as critics of 'Fortress Europe'.

Overshadowing all these controversies is the so-called 'normalisation' of Christiania. It had been operating outside of Denmark's legal framework for forty years when, in February 2011, the government proposed a 'take it or leave it' deal in which residents were asked to buy the land and buildings or risk being forced out. Faced with the biggest challenge in its history, Christiania closed its public entrances and engaged in four days of debate. Divisions ran deep. Some argued for complete normalisation, allowing people to buy their own homes. Others said the deal should be rejected, that it signalled the end of utopia and – this being Christiania – expressed their frustration through interpretative dance. In the end the deal was accepted: a collective purchase was agreed. Supporters of the deal, such as Emmerik, said Christiania was able to 'buy itself free of speculation'; the experiment could continue, Christiania was safe. Yet there is no getting round the fact that Christiania was being mainstreamed: forced to accept – on paper, at least – the same planning regulations and building controls as the rest of the country. On 1 July

2012 the Christiania Foundation was created, to purchase the land and buildings.

Despite this tryst with compromise, Christiania still inspires hope. The desire to live without rules and regulations, to be able to express oneself freely, is not just a whim here; it is a steely determination. For all its faults and contradictions, it matters that Christiania survives. This 'escaped zone' shouldn't be claimed to be perfect, then condemned for failing to live up to that fantasy. It is a work in progress, an alternative island that has a rough-and-ready relationship with the outside world. It shows how a collective way of living that is also highly individualistic can thrive in, and contribute to, the modern city.

Helsinki Wild Harvest

The small yellow berries are soft and overripe. I try to tug them gently from the thorny branch, but they disintegrate immediately, gushing copious thin juice over my frozen fingers. Even so I manage to get two of the waxy, oval fruits of the sea buckthorn bush to my lips, and, having been to Finland before and got a little used to the medicinal tang of this widely drunk wonder berry, I relish my prize. I'm even counting my foray into urban foraging a great success – an unexpected one, since I'm picking them at the wrong time of year and in the middle of a small waterside park in the heart of the country's capital. It's deep into December, and even at midday it's minus 6 degrees. But my printout of the online 'Helsinki Harvest Map' has led me straight to this tall bush. Like all the other entries, it has been given a unique biography (though I'm rather hurt to find my hard-won find is written up as 'Old bush. Very low yield and small berries').

135

The pursuit of wild food is the pursuit of an ideal; an attempt to reimagine the city as a garden. This other Eden takes work to uncover; it does not fall readily into one's hand. But that's the point. It demands that we look at the city sideways; seeing not the concrete and bricks, but nature's bounty. I'm finding the Helsinki Harvest Map to be an addictive means of finding my way around even when, as now, a bone-cutting wind is blowing straight off the Gulf of Finland. There is no one else about. Or so I think. Clambering up behind some pines onto rocky ground in order to meet up with an apple tree ('Small, old tree with dozens of apples varying in size, can be eaten but better for jam') a thin hare, mottled with a half-formed white winter coat, bolts out in front of me and stands rigid, stretching up on its hind legs to gain a better view. Coming from England, where hares are now a rarity even in the countryside, it's electrifying. I'm within a few metres of a mountain hare in a small city park. Clearly I'm not the only creature trying to gain sustenance from this blasted landscape.

My competitor has such a starved look that my efforts at playing the wild man suddenly feel a little hollow. The truth is that I don't need to make my meals from wild berries. But it is such fun. The Finns are supposed to be taciturn: it's easy to spot an extrovert Finn, I've been told; they're the one starring at someone else's shoes. However, even the Finns get animated when it comes to wild harvesting. I'm here for an academic conference – it has the ambitious title 'The West: Concept, Narrative and Politics' – and it's being held in a modern, unassuming city in the middle of the country called Jyväskylä. In between chin-stroking about the enigmatic qualities of the West, I buttonhole Jukka and Henna-Riikka, the youthful and charming conference organisers, about their foraging habits. It turns out that they, like most Finns, have enjoyed wild harvesting from childhood. I can tell they're a bit surprised by my questions: *of course* they go out foraging; doesn't everyone? There are thirty-seven types of edible berry growing in Finland, and their names are a beautiful thing, a roll call of sweet moments: lingonberries, bilberries, cloudberries, raspberries, cranberries, arctic brambles, wild strawberries, bog whortleberries, mountain crowberries and rowanberries, as well as sea buckthorn berries. Then there are nuts, mushrooms, syrup from birch trees, flour from pine bark, all sorts of herbs and plenty of hunting and fishing.

Reconnecting the city with nature is one of the most important and pressing ideas of the twenty-first century. It's also one of the most practical utopias. Like most places in Finland, Jyväskylä is surrounded by an endless horizon of pine forests and, at this time of year, frozen lakes. Finns are rightly proud of their legally enshrined 'everyman's rights': everyone has the right to roam, camp and gather wild food everywhere apart from in private gardens. This is a country with deeply held egalitarian instincts: nature's bounty is seen as a common heritage. Perhaps it is also worth noting that in a country of just five and half million, there are 632,000 individual family forest owners. The assumption that 'urban living' equals a disconnection from nature, or that city folk suffer from 'nature deficit disorder', doesn't seem to apply here. Ordinary towns and cities like Jyväskylä are often overlooked when we talk about city life. The urban has come to be defined in terms of massive metropolises, places like New York or London. So we forget that most urbanites live in much smaller places; many in places like Jyväskylä, where you can see the forest from most street corners and where a 'bond with nature' isn't an affectation but something everyday and undeniable.

That's how it was once for me, too, or so I like to think. In my childhood, prizing out sweet chestnuts and picking crab apples felt like common things, to me at least. It was blackberry picking that defined late summer; that turned it into 'blackberry time'. Reaching for the fattest, furthest fruit, proudly carrying the sagging plastic bag of berries homeward; what could be better? So it doesn't surprise me that so many people are trying to rekindle such simple delights and are remapping the city as a network not of roads and buildings, but berry bushes and apple trees.

You have to head to a bigger, more fashionable city, to fly down from Jyväskylä to Helsinki, in order to see wild harvesting turned into something bohemian and chic. Indeed, what is now badged the 'foraging trend' is widely attributed to the world's most cutting-edge and stylish restaurant, Copenhagen's Noma. Here the taster menu, including 'first apple of the new season' and 'deep-fried reindeer moss', cost more than £200. Noma closed at the end of 2016; its owners intend to create an 'urban farm' near Christiania.

Urban foraging has both an epicurean and an environmentalist face. The latter is more to the fore in the London 'wild food scene', which has found its bible in John Rensten's *The Edible City: A Year of Wild Food*. London groups like Hackney Harvest and Urban Harvest organise edible flower walks, and are dedicated to mapping out every last fruit tree and edible weed patch. London's community has kin in many parts of the world: Melbourne, Sydney and New York all have active networks. In San Francisco a group called ForageSF support an Underground Market and a nomadic supper club, the 'Wild Kitchen'.

The magic of conjuring food from the city appeals to the modern taste for the eco-friendly, the locally sourced and the outré. A collision of affluence and thrift, primitivism and sophistication makes for some paradoxical outcomes. The blurb written for one Cape Town restaurant runs: 'After a hard day's foraging work, the luxury Table Bay Hotel is a great sanctuary to retreat back to, with waterfront views and Table Mountain as a backdrop'. Or how about downtown Manhattan's luxury Peninsula Hotel, where guests can enjoy wild picnics and even be 'taken to hidden spots around Central Park where herbs and vegetables grow in the wild'? The simplest pleasures are being sold back to us as opulence. Flying out to Jyväskylä, I read in my Finnair in-flight magazine that the country is now popular with 'a growing contingent of wealthy A-listers willing to pay a premium for indulgences such as fresh air, silence, pristine nature and green luxury'.

Such is the volume of well-heeled pluckers and pickers that, according to one London wild food site, their activities are 'being banned by some parks, who fear they will be pillaged by mushroom hunters supplying fancy restaurants'. But the new trend for urban foraging is just a small expression of something much wider, far older and much more everyday. People in cities and towns all over Finland have had wild harvesting maps in their heads for centuries, and they will carry on using the city in this way long after avant-garde restaurants get bored with serving hand-picked weeds to their customers. In places like Jyväskylä, where A-listers might feel as welcome as a Chicken McNugget on a Noma tasting menu, the town and the countryside have never been that separate.

Back in Helsinki, I'm trudging up through the ice-buried streets to see if I can find a particular walnut tree on a patch of rough land overlooking the busy ferry port ('A large tree. Nuts are falling on the wrong side of the railing, but there's a lot'). It's not easy: all the leaves that might have helped me find it have long gone, and I am frozen to the bone. Eventually I kick up some wizened nut husks and soon find that the ground is matted with them. I'm satisfied that I've found my tree. It's an unregarded giant. It's barely glanced at, and goes unnoticed amid the noise of lorries grinding off and on the ferries, but it's magnificent. A few months ago it was plentiful. We all know that it will be again. Come next autumn, there will be another harvest.

City of
Helicopters

French engineer Paul Cornu was the first to ride in a helicopter. In 1907 his whirling, four-bladed contraption
managed to lift off with him inside. Cornu only achieved a
vertical ascent of one foot, but it was enough to turn helicopters from a fantasy into a reality. The following decades saw
the helicopter ascend ever higher, to become one of the icons
of mobility and modernity.

The Brazilian metropolis of São Paulo is the closest we have
got, so far, to Helicopter City. At least four helicopters land or
take off in São Paulo every five minutes. It's the place to go to
see how the helicopter changes things and for whom. São Paulo
is a megacity; an urban behemoth of some 22 million people
spread over a subtropical plateau. The title of 'world's worst
traffic' is hotly contested, but São Paulo is a favoured nominee.
Traffic jams here have been recorded snaking a colossal 183

miles. It takes hours for ordinary commuters to get into and away from work. To add to the misery, once they are stuck in traffic they become sitting ducks for nifty thieves – all those immobile people are tempting targets. Worse crimes are also common, including murder and kidnap. Kidnapping has become so common in São Paulo that some plastic surgeons specialise in reconstructing parts of the body, like the ear, prized by kidnappers as proof of their bounty.

Security is big business in this city, with a flourishing trade in electrified fences, bulletproof cars, gated communities and, of course, the ultimate vehicle for escape, the helicopter. São Paulo has 50 per cent more helipads than the UK. That may sound like a lot, but urban helicopters are pretty tiny. They cater for the few, not the many. In his amazing book on the literal ascent of the wealthy, called simply *Vertical*, Stephen Graham (who's a friend and neighbour of mine) has worked out that if São Paulo's helicopters take on average three passengers, 'this aerial transportation system can shuttle around only three-quarters of one hundredth of one per cent of São Paulo's population at a time'.

This is a utopia for the rich: Helicopter City extends and completes their enclave. It takes their ability to live high above, and move rapidly away from, the rest of us to a new level. Typically, these helicopters fly from gated communities on the outskirts of the city to its central Business District, or between the rooftops of hotels and the main airports. What is sometimes called 'successionary affluence' is not new: escaping the masses is as ancient as wealth itself. But the helicopter brings a new dimension to this kind of spatial segregation. The helicopter owners are free to roam in the city's three-dimensional space while the hoi polloi are stuck travelling along prescribed and linear routes and, for the most part, grounded in 2D.

One helipad installer talking on a São Paulo TV station in 2005 suggested that they had become a de rigueur embellishment for any affluent high-rise. 'Today we define the sculptural rooftop helipad as a good architectural result,' he explained. 'It works like a sculpture.' Brazilian sociologist Saulo B. Cwerner, who has studied the impact of the helicopter on São Paulo, tells us that 'many new buildings are designed with helipads from scratch. Although increasingly regulated, the design of rooftop helipads still allows for expressions of corporate identity and power, especially through the use of metallic structures.' Cwerner is interested in the helipads, and in the swarms of choppers they attract, as visual symbols; high-profile displays of dominance and power. It is a 'highly conspicuous mode of arrival and departure', he writes, especially given the enormous noise that is created. Members of the helicopter class have striven to detach themselves from the street, but their way of doing so sends a clear signal: you're down there and we're up here.

Stephen Graham calls it 'elite helicopter urbanism', and argues that it 'works to sustain and integrate a broader process of vertical succession of the super-elite, who gravitate upwards in their living, working and leisure environments'. A separation of levels has occurred, turning the street into an object of fear and loathing for 'elites that now ascend over, and cease to rely on, the city's chronically saturated ground-level streetscapes'.

The noise of all these helicopters has meant that since 2009, restrictions have been placed on the siting of helipads near schools and hospitals and their freedom to overfly residential areas. São Paulo has the world's only dedicated helicopter control facility. It tries to direct them along a fixed grid. The map of helicopter routes shows narrow, 660-foot-wide lanes above the rivers, large avenues, highways and railways. However, helicopter pilots don't always follow the rules and regularly take alternative pathways, nipping over the rooftops of neighbourhoods at all times of day, much to the annoyance of the people living below.

It wasn't supposed to be like this. For a while the helicopter was going to be part of everyone's future. 'Will There Be a Plane in Every Garage?' asked one American pamphlet from 1945. The answer seemed clear: 'If your home is in a suburban or rural district, the helicopter can take you to and from work daily in comfort and with speed. You won't get tied up in a traffic jam or have to stop for red lights.' The 1960s cartoon series *The Jetsons* captured the mood. Living in Skypad Apartments, the Jetson family zipped between odd-shaped domes balanced on high-rise stilts; nothing ever seemed to touch the ground.

But the anticipated roll-out of helicopter travel to ordinary families, or even fairly well-off ones, never happened. The most obvious reason is that helicopters are expensive. But this used to be true of cars. With the same kind of mass manufacture, they could be much cheaper. Other explanations are that helicopters require skilled pilots, they need landing room and they're horribly noisy. But, again, these could all be overcome.

It's a mistake to imagine that the helicopter age didn't arrive because it couldn't. It's probably closer to the truth to say that the challenges were bigger than anticipated, but the story is not over yet. As the world's traffic gets worse and worse, I suspect that the attraction of the helicopter will only grow. The helicopter age is still on the horizon.

That's why São Paulo is such an important place. It's ahead of the curve. And it doesn't only teach us that Helicopter City is a plaything of the super-rich. São Paulo, like the rest of Brazil, has experienced an economic crash. There has been a 20 per cent decline in helicopter flights, and as a consequence operators are reaching out to new markets. Uber began to roll out a helicopter service in São Paulo in 2016. It announced that customers could catch rides between four airports and five heliports around the city for just US$20 (about £15). That was a promotional price; the normal fee is touted at US$63 (£49) – quite a lot, but within the reach of a whole new swathe of customers. Writing for *Bloomberg News*, Blake Schmidt has described his experience of the Ubercopter. 'An UberX driver wearing black sunglasses picked me up,' reports Schmidt. Once at the heliport, 'It took about 15 minutes for clearance, a minor drawback for someone on deadline but not nearly as agonizing as São Paulo traffic.' The flight itself took only 10 minutes and cost 'about twice what the trip would have cost by cab alone. The same trip in a cab, however, could have taken an hour or two in rush-hour traffic.'

Unless the problems of other modes of transport get sorted out, the twenty-first century is likely to see a return of the dream of helicopter cities. There has been an expansion of helicopter provision across India and China. The lure of leaping over the traffic, of moving from two dimensions into three, grows in appeal in direct proportion to the dysfunctionality of surface transit.

City without Ground

Maps are flat. They represent space as two-dimensional. In real cities, roads fly over each other; people are stacked in layers; pathways, escalators and lifts are arranged in three-dimensional space. In some cities our oldest reference point, the ground, is disappearing from view altogether.

Cities Without Ground: A Hong Kong Guidebook is the world's first three-dimensional map book. Each page offers a different brightly coloured and geometric multidimensional map, with 'automated people movers', terraces and walkways leaping between the transparent vertical outlines of skyscrapers. Its authors, Adam Frampton, Jonathan Solomon and Clara Wong, claim that Hong Kong is 'a template for public space within future cities undergoing intense densification'.

It looks alluring, but having shown it to people who regularly have to navigate Hong Kong, I'm not sure it's that practical.

My colleague, Dr Michael Richardson, who every year leads geography students all over the city, turns the book over in his hands, initial enthusiasm rapidly turning into a perplexed frown. 'This would get me even more lost,' he concludes. When I ask him about the book's upbeat message, its claim that Hong Kong is a beacon for other places because it shows that 'public space does not require stable ground', he counters with a more pragmatic observation: 'It's impossible to cross the roads! You have to double back and keep going up into shopping spaces to get anywhere.' For Michael, Hong Kong's 'city without ground' is 'all about getting you into those commercial areas'.

Perhaps, rather than trying to break free from the ground, we should be trying, even in densely packed cities, to have a closer connection to it. Is it a good idea to lose that reference point, that link to the earth? I doubt it. But I seem to be pushing against the tide; the buzz at the moment is all about the pleasures of being lifted up and letting go. When the anthropologist Tim Choy describes a Hong Kong executive who 'wends his way expertly through Wanchai, a government and night-life district on Hong Kong Island, without ever touching the ground', it sounds not just fun but something to aspire to. Hong Kong is estimated to have 60,356 lifts. But the most iconic element in its groundless infrastructure is its escalators. The Central–Mid-Levels escalators in Hong Kong comprise the longest outdoor covered escalator system in the world, although its movement of direction depends on whether it is ferrying commuters to or from work: from 6 a.m. to 10 a.m. the escalator moves downhill, and from 10.15 a.m. to 12 a.m., uphill. Expanded to connect to shopping malls and the port, people go to work by escalator as well as go shopping or visit restaurants, all without ever setting foot on solid ground or leaving a covered area.

Hong Kong's vertical layering emerged piecemeal, with no real overall planning. It was linked together into a network more by dint of the density of the connections than because of oversight. Nevertheless, it is now being offered as a model which other cities should emulate. The promise of ultra-density, of having a functioning city that doesn't sprawl out but is highly focused, an intense bright point of innovation and activity, has become irresistible to twenty-first-century planners. Adam Frampton explains that 'people are thinking a lot about how to emulate some of the patterns you see in Hong Kong, simply because it's not viable for everyone to follow the American model of having a car and driving from work to home'. After *Cities Without Ground*, Frampton began working on plans for a new zone within the Chinese city of Shenzhen, where the civic authorities are also striving for groundlessness. He notes that the planning brief 'stipulates that they want public levels at 0 meters, 12 meters, 24 meters, and 50 meters', thus creating 'layered public space in a way that is modelled after Hong Kong'.

Three-dimensional plans are not new. Leonardo da Vinci's notebooks include a plan for an underground canal and a waterwheel to power fountains – all in 3D. Architects have used similar 'see through' designs ever since. But such plans are seen as technical drawings and are restricted to particular buildings. They don't range across the city and incorporate whole neighbourhoods. There is, or perhaps I should say there was, another and more obvious difference between plans and maps: one is a design tool, the other a guide for getting about. It's a difference that may be starting to crumble as 'cities without ground' rise ever further from an increasingly indiscernible earth. As maps morph into plans, the feeling that we live in one giant construction site – that nothing will last – will only get more powerful. Such three-dimensional portraits offer a new kind of cartography, one designed for the needs of rapid change and continuous building.

However, an architectural colleague of Frampton, James Schrader, makes a telling comparison: in 'a lot of Western cities', he notes, 'the skywalk systems of the 1960s and 1970s were, a few decades later, largely derided as failures that killed the life of the street below. Whereas in Asian cities, it's so dense that layering actually works' (see the chapter 'Skywalks'). It seems that the errors of the West are being reborn as great successes in East Asia. There is certainly an appetite among architects to believe so, and for proclaiming the drama and excitement of a new kind of city that is fully post-natural, where the ground has been made redundant. Frampton, Solomon and Wong contend that in Hong Kong, which has seen a lot of land reclaimed from the sea, the 'physical ground is equal parts elusive and irrelevant. What appears to be terra firma was likely water not so long ago.'

At the end of their guidebook, Frampton, Solomon and Wong include a short but telling chapter that hints at the environmental consequences of cities without ground. Called 'Atmosphere', it looks at the different levels of air-conditioning across the city: the fans that keep the hotpot restaurant at 22.9

degrees and the train carriages a chilly 20.5 degrees. Chugging with air con, the city without natural ground is also the city without natural temperatures. Environmentalists have long tried to steer the city away from the idea that, as the 'Go Green Hong Kong' (gogreenhongkong.com) website has it, 'we have transitioned to a post-soil society, where with enough concrete and wifi all of our needs can be met'. One would hardly believe it to be true from the commentary of many of the architects who thrill to the city's groundlessness, but there is an acute sense of environmental loss in Hong Kong, albeit a frustrated one. One academic study of public attitudes notes that the 'majority of the public in Hong Kong have repeatedly told pollsters that they believe environmental problems have become very serious in the city', but also that they are 'highly sceptical about the actual degree of concern felt by their neighbours, friends and relatives'.

A disconnect with the ground isn't something to be celebrated, but rather to be mourned. Jan Gehl, an architect from a different tradition to the authors of *Cities Without Ground*, even tries to argue that anybody living above the fifth floor is in trouble. To be a high-liver means that you are 'not part of the earth anymore, because you can't see what's going on on the ground and the people on the ground can't see where you are'. Gehl isn't just talking about the need for human-scale cityscapes but, more fundamentally, cities that understand that human beings and all their works are part of nature and must never forget it. Perhaps Gehl would find some grim symbolism in the fact that in 2017, eighteen people were injured when one of Hong Kong's escalators suddenly went into reverse.

Richard Louv, who coined the phrase 'nature deficit disorder', argues that 'the future will belong to the nature-smart'; people and communities who have learnt to 'balance the virtual with the real'. For 'the more high-tech we become, the more nature we need'.

Many ordinary people in Hong Kong, and even in the city's government, seem to agree. The government has planted 87 million trees and shrubs since 2006, and has a comprehensive greening policy. One small but telling part of this policy is the annual 'People, Trees, Harmony' photography competition. I found looking at the winning entrants, nearly all of majestic, broad-beamed trees, surprisingly moving. A little later it dawned on me why: what I was looking at were images that pay nostalgic homage to something that matters to us all, something we all revere – the ground.

IV.
Ghostly Places

All places have ghosts. I'm not thinking about haunted houses or creaky old villas on the hill, but about ordinary places: shopping centres, car parks, metro systems. Even when you erase the past, wipe clean the landscape, something always remains; a sense of unease, a kind of modern chill. That's what blows down the tunnels of Tokyo's Shinjuku Station and, in a pathetic and far more dilapidated way, along the narrow spines of the once-modern Skywalks of Newcastle. Other types of ghosts can be found if you pick your way across the rubble into the Boys Village in Wales, or down the winding lanes of the Himalayan town of Shimla to a British graveyard, where the last resting places of the country's former rulers are grown wild and presided over by a leopard. Some ghosts summon you, but others are called forth; so it is with the extraordinary recreation of a Soviet-era city fashioned by the director of *Dau*, perhaps the strangest film story of all time; all the more so since, years after it starting shooting, there is still no film. Also summoned into being are the mythological, spell-woven places of Magical London, although by comparison these magicians are eminently sensible; they tread lightly in order to listen to the city's enchantments. We should, of course, have listened to the Tsunami Stones: dotted along Japan's northern coast, they warn of the danger of building near the shore. Erected centuries ago, their message was ignored. Can one create places, such as Nuclear Markers, that warn of radioactive waste – that send messages to the future? How will we haunt people in tens of thousands of years' time to warn them of danger, when our language, our culture, will be long forgotten?

The Phantom Tunnel
of Shinjuku Station

My aim is to get lost in the busiest train station in the world. About four million people a day pass through the countless levels, 200 exits and 36 platforms of Shinjuku Station in the heart of Tokyo. It is a temple of efficiency and signage, but there is a legend attached to it. It's not a legend from days of yore; it's a modern legend that reflects the fears and fantasies that come to life in an overwhelming metropolis.

Sometimes called the Bermuda Triangle of Tokyo, the story goes that some commuters never make it home. They take a wrong turn, then another, get flustered, run down the wrong stairs and end up in the wrong elevator, until they find themselves quite alone in a quiet corridor, the soft boom of a distant underground train sounding somewhere far above them. They are never seen again.

There are no verified cases, no roll call of the disappeared. But it's a recurring story which touches on something profoundly unsettling about our relationship with the city. Tokyo is so big, so impersonal and machine-like that it is not hard to imagine that we could disappear inside it, that it could eat us up: that it contains phantom limbs, unnamed and hidden tunnels and lanes into which the unwary are drawn and where they are swallowed up.

On a hot and humid day in late August I entered Shinjuku's maw. It's not rush hour, but people are pouring through the broad, immaculately clean and well-signed corridors. I thought I'd need to get through the ticket barriers to go into the belly of this thing, but after some toing and froing through a few of them, with the assistance of some puzzled staff (the barriers here open to let you in, but if you haven't gone anywhere they won't let you out), I'm finding that the loneliest spots are in the meeting ground between the station and the giant shopping complex that engulfs it. For all its many exits and corridors, the station is just one element of a much bigger labyrinth. It is in this interzone where the crowds thin out and I am able to worm my way down deeper and deeper, heading into the furthest and quietest corners of the labyrinth. It takes about twenty minutes, down an escalator, down some more stairs and another escalator, before I'm anywhere that isn't a major thoroughfare. It's a stairwell with no signage and I head down again; and here the noises of the crowd have softened to almost nothing. It's a beguiling contrast; the idea of ever returning up there, I suddenly realise, is deeply unappealing. It's so nice to hear the clip of my own footsteps. The lighting here is not so bright, the cleaning regime less evident; there are even a few items of litter, blown down from above, and a man is slumped against a huge paper shopping bag on the flight below me. He's very still, his unseeing eyes focused on nothing.

It makes sense that he's down here. The Japanese don't tend to go in for public displays of angst or even grumpiness; smiling is almost compulsory. It is only at the edge of things, such as down here, that misery is possible. Shinjuku is the home of coteries of ghosts. Not all of them are scary; one group of spectres has taken on the helpful task of saving people from suicide. They are said to push suicidal passengers away from the tracks to safety. These useful ghosts were the unfortunate victims of a secret mass suicide and they now drift about, making sure others don't suffer the same fate.

Deep below the city's roar there's a kind of camaraderie of the fallen. I'm beginning to understand how, if I'm not careful, if I don't watch my step, I could get ensnared. It's lulling, if a little unnerving. Something is replaying in my memory; something frightening. Late at night, maybe forty years ago, I saw a short Spanish film on the TV that has stuck fast. It is called *The Telephone Box*, and starts with a group of workmen installing a red telephone box at a busy intersection in Madrid. It is usually classed as a horror film, and, though it doesn't fall easily into any category, it has a dislocated atmosphere. There is no dialogue, just the kind of murmur and machine noises that you hear all around you in Shinjuku. A suited middle-aged man goes in to use the phone, but when he tries to leave he finds the door is stuck. A comedy of helpful passers-by, of police and firemen pushing and straining, ensues. It's funny;

the amused helpers and onlookers and the stoic embarrassment of the trapped man. Telephone engineers arrive, and
there commences a bizarre journey on the back of their flatbed
van, with the man still in his box, always completely visible,
being driven away. The shift from comedy to the macabre
is completed as the box is deposited in a vast cavern, the
resting places of numerous other phone boxes, each of which
contains decayed and skeletal remains. Of course; you saw it
coming. You always knew. The horror is the horror of banal
modernity and, more than that, the awakening of a commonly
shared sense that the city is impossible, beyond us; that we slip
between its spaces like ghosts.

Getting lost confronts us with our own insubstantiality, but
it is a difficult thing to do on purpose. I don't know quite where
I am or how many levels into the earth I have come, but rather
than the panic and paranoia that are the accompaniments of
the truly lost, I'm getting sleepy. It's so warm down here; quiet,
too; just me and the slumped man and, below him, a series of
metal-panelled doors leading off from each turn of the stairwell. They are all clamped shut, though some are hissing with
malevolent force.

Some people believe there is a secret city under Tokyo.
Japanese journalist Shun Akiba claims to have come across a
map in a second-hand bookshop that reveals secret parallel
railway lines and many mysterious tunnels. In his book
Imperial City Tokyo: Secret of a Hidden Underground Network,
which has run to many editions, Akiba tells us that this knowledge is being suppressed. When he first tried to break the news
he found no one wanted to know, their 'lips zipped tight'. He

woke up one day, he says, to find his thighs glued together into a jelly-like mass. 'Subway officials treat me as if I'm a drunk or a madman,' he complained. Akiba thinks that the hidden network was built before the Second World War and is now part of the country's preparations for a nuclear attack. What is now labelled the 'Tokyo underground secret route theory' joins a library of similar conjectures about almost every other major city in the world. The content of these theories is less interesting than the need to keep on producing them. The idea that there is a shadow city, a city under the city, seems irresistible. Maybe to set out to find it misses the point. For the city's landscape is imagined to be, in some way, always beyond us; it's true nature endlessly receding.

And the ghosts are us: that is especially palpable amid the endless human movement of places like Shinjuku. I've walked myself into a kind of trance, but it's a familiar, almost comforting feeling, of leaving no trace, remembering no faces. Even though I stand out as a foreigner, the visibility of one's public presence in such a busy place is actually a kind of invisibility; everyone sees you, but you're a nothing; you could be anyone. I'm beginning to think the yearning for the phantom tunnel may be about wishing into being a landscape that mirrors our own condition. It's a strange lure; after all, I started this venture with metaphors of being eaten up by the city, of falling into its jaws. I'm getting lost down here in more ways than one; and I'm not the only one. Emerging out onto a glass-fronted terrace that affords a fine view of numerous tracks, carrying both colour-coded local subway carriages and sleek bullet trains, I'm trying to weigh up what just happened to me.

There are families playing with giant cartoon penguin figures, and a German beer stall, and everyone is happy. It is unbearable. I find some seating around a corner behind helpfully large planting. And here's a man, well dressed, early forties, head in his hands, who I think is crying though he's making no noise. It occurs to me that, far from just happening upon him by chance, I have searched him out. What can we do, we two? I sit away from him, not knowing where to go. This started as a rather flippant quest; I went looking for the places where people had disappeared and been consumed by the city. I'm not sure I really wanted to find them, but I did.

Skywalks

The modern city isn't new any more; it's old and layered with yesterday's bright plans for the future. Visions from the past, utopian master plans and heady schemes from long ago are piled one on top of the other, and we are left picking our way through the pieces.

The Skywalks are a remnant of one such dead hope. They are laced across my home city of Newcastle upon Tyne in England's North East. Built nearly half a century ago, they were just one element in what was supposed to be a complete refashioning of the ancient city of Newcastle into the 'Brasilia of the North'. The promise was that motorways would plough underneath, and high skywalks and skydecks for pedestrians would glide above.

The whole system was never completed, but bits of it made it into concrete; a fat stub of motorway that bisects the city then suddenly stops and turns into ordinary roads, and the long, disconnected strands of skywalk. The utopian presence

163

in the modern city is often like this; fragmentary and leftover. The Skywalks have a certain modernist elegance, but they go from nowhere to nowhere. In any case, it's not their design or planning that fascinates me, but how we live with them today; making our way through the debris of a forgotten vision of tomorrow.

The most famous examples of utopian remnants echo much darker pasts. A 66-foot-high concrete cylinder is one of the few reminders of Hitler's plan to remodel Berlin as 'Germania'. It was used as a heavy-load test structure, built to gauge the weight of one of the four pillars of a 390-foot-high Triumphal Arch destined to tower over the new city. Berlin is self-conscious about these vestiges, unlike most cities, such as Rome, where Mussolini's own hubristic plan is visible in the form of a grand road. The Via dei Fori Imperiali, which spears its way between the Piazza Venezia and the Colosseum, is an incongruously wide highway. It was built for fascist marches, and destroyed a wide swathe of the ancient city. Mussolini also appears in Rome's Foro Italico sports complex (once called the Foro Mussolini), which is still presided over by an angular obelisk proclaiming 'DUX' (the Latin for Duce, or leader) and boasts the largest mosaic plaza built since the fall of Rome, emblazoned with 248 iterations of 'IL DUCE'.

Plans for total urban transformation were just as popular with left-wing as with right-wing regimes, and post-war European welfare states subscribed to the same basic notion: that the future demanded the eradication of the past. In some ways these plans were even bolder than that which went before, for they eschewed classical references in favour of stark brutalism. These post-war visions were not communicated by fiery oratory, but in the dreary language of efficient access to shopping and easy commuting. The architect responsible for the Skywalks was Wilfred Burns, an affable and earnest visionary who had come to Newcastle after overseeing the dramatic post-war redevelopment of Coventry. In his 1967 *A Study in Re-planning at Newcastle Upon Tyne*, he muses 'whether vehicles should be on top of the pedestrians' before opting for the other way round; a solution that allowed what he called the 'segregation of pedestrians and vehicles in the vertical plane'. Given the scale of the rebuild, his overall aim was shockingly mundane: to 'provide for easy servicing to all the shops and to provide car parking facilities with access to cars at the road level but egress by pedestrians on to the shopping deck'.

Wilfred Burns' language was prosaic but, if completed, his master plan would have allowed people to walk the length of the entire city centre without having to cross a single road. The real winners, though, were to be the car drivers, who could hurtle through the city without touching their brakes. Burns' career progressed rapidly, and in 1980 he was awarded a knighthood. Conversely, the pugnacious leader of Newcastle city council who oversaw his plans, T. Dan Smith, was convicted of corruption in 1974 and spent three years in jail.

Walking the Skywalks is an oddly melancholic experience. There are a few redundant clues to the complete scheme: odd blocks poke out of the walls on some city centre buildings, structural ledges intended to support a skywalk that never arrived. The skywalks that did get built are not just bypassed by most people; they are literally off the map. Despite being touted as key access routes, the *A–Z* of Newcastle doesn't bother to include them. You have to live in the city for some time before you work out that they are of any use. It takes a further few years of use to gain confidence on them because they dart off in odd directions and will blindside any novice. It took me twenty years before I got the hang of them.

That happened by chance, when I came across a map online, a hand-drawn annotation of an aerial photo with the Skywalks helpfully coloured in. It was part of a series of images on the 1960s master plan, complete with models of the megastructure the Skywalks were designed to lead to, the Tyne Deck. The deck would have sat over the River Tyne and been a 'symbol of rebirth'; the platform for great 'public buildings'. Its architects proclaimed that it would overturn 'reactionary parochial attitudes'. It would also have turned the Tyne from a river into a 'linear lake'; a vestigial puddle that would have obliterated the very thing that the city of Newcastle upon Tyne is named after.

Finding that map of the Skywalks opened up a hidden trail; a passageway into the unseen city. I began to walk the route and take photographs, preparatory to an expedition for a student group. It's along that preparatory walk that I will take you now. It's one I value because, although I use the Skywalks every week, they are not what they were. The city council has been blocking off key sections and pulling down small bits. It doesn't seem to know what to do with them, but it is clear the Skywalks' days are numbered.

I start at a broad pedestrian ramp that narrows into a thin, stilted walkway from which you can see several other skywalks seemingly flung between random points. It's a steep climb, and it's hard to imagine any group of pedestrians who'd not find it quicker to move about at street level. It's empty, as usual. The broken glass and beer cans that roll noisily in the stiff wind will be up here for months.

It's breezy on this high strip, but it soon culverts into a smelly, high-rise alley running between blankly modern brick buildings. It's another half-baked plan: a hotel and leisure complex that must have seemed like its own form of utopian space back in the 1980s but has long since turned into somewhere scuzzy and unsafe. Emerging onto a high stairwell, I can see on one side the site of what was once a large cinema complex. It was pulled down and today is a shiny, thousand-windowed university campus. But up here everything feels temporary. How long will these high lanes, that campus or indeed any of this, last? It's barely worth storing memories for such places because none of them feels like it has any claim or a real presence on the earth. Like me, they are just passing through. From this same vantage point I can see another example of an ephemeral place, the Blue Square, an artist-built pedestrian surface composed of what was supposed to be shining blue tiles, finished off by neon-lit, surface-level glass panels. The panels stopped working soon after installation and, as if mourning the loss, the tiles of the Blue Square turned grey.

The Skywalks weave their way around and between these various layers like an infestation. It's a disorientating journey. I thought I knew them well, but I'm getting lost; there are too many small levels and turns and, though I'm in the middle of a large city, there's no one up here; no one about to ask the way.

I'm headed for the secret heart of the maze. Down near the quayside the Skywalks turn in a ziggurat of multiple staircases before doing something deeply odd. On the left side, one route leads you down past a flagged square. The square has a nightmarish quality; it's not just empty but walled away; surrounded on three sides by offices and, on its fourth, by a low brick wall. It's a bizarre anti-space that is utterly hidden from the busy shopping street that is just metres away. You have to climb the low wall to get into it and, for reasons I can't quite fathom, it feels like one of the saddest places on the planet. There's another Skywalk route that doesn't dip down to this hopeless square but just keeps on going, driving a pathway on huge stilts under the massive Tyne Bridge that soars above. Then it just stops, suspended under the bridge; unvisited and unusable.

Like so many others, I've spent a lifetime trying to navigate the fragments of other people's grand plans. The original graphics for the Skywalks were populated by smart-looking, generic nobodies; blank-faced nothings. Whose utopia was it, I wonder? Not mine. In those nobodies I see myself, all of us, unknown and nameless but still walking, crossing roads, carrying on, having to find our way through somebody else's vision of the future.

The Boys Village

I don't see my older brother Paul often; three times a year, maybe. He lives in Glyncorrwg, a hard-up ex-mining village at the end of a long one-way road in one of the valleys of South Wales. The last time I was there, for a few days in the spring of 2016, some species of nostalgia must have been at work. Just looking for somewhere interesting on the map to drive to, we chanced upon an intriguing place name, tucked behind the coast: 'Boys Village'.

It's an abandoned holiday camp set up in 1925 for boys from the Welsh coalfields. It was built to last. Despite its derelict, shattered condition it is still an impressive complex, and ranges over many acres. The solid church with a tower and vaulted hall is surrounded by restaurant halls, a sports hall set up for basketball and tennis, an empty, weed-infested swimming pool and numerous residential blocks whose glassless windows afford views over quiet fields. It looks less like a spot for week-long vacations than a citadel built by a tribe of sport lovers; or a place that has befallen a sudden calamity, the townsfolk fleeing one night, leaving everything.

The Boys Village has a strong atmosphere; more so than many of the other abandoned institutions I've wandered into. All those ruined hospitals, power stations and tunnels are exciting, but they were work places; people didn't go there for fun, and they still feel machine-like and oppressive. This lost village is different. Perhaps it's wishful thinking, but I'm getting the feeling that this was a happy place; open and jolly. I come across some original wallpaper, a bright orange patterned strip flapping in a shattered bedroom. The ceiling has caved in, but you can almost hear the excited shouts, the boys dashing out of the door, off to the pool or the beach; that same door which now lies heavily in a corner of the room.

The village's biggest open area, in front of the church, was a parade ground. It is now broken up and scattered with rubbish, but a stone war memorial stands resolutely to attention at its centre. Some residual respect seems to have stayed the hand of the vandals; it hasn't been sprayed by graffiti tags. I wrote down the inscription.

DEDICATED

TO THE MEMORY OF

THE YOUTH OF ALL

NATIONS WHO FELL

THAT WAR MIGHT END

BY THE BOYS OF THE

SOUTH WALES COALFIELD

WHO AT THIS ALTAR

DEDICATE THEMSELVES
TO COMPLETE THE TASK
SO NOBLY BEGUN

Every year an old miner and his wife would be chosen as 'Lord Mayor' and 'Lady Mayoress' of the camp and given a free holiday in a specially built cottage. There are photos of them in *The St Athan's Camper*, the village's magazine: old couples – he in a black suit, she in a smart long dress – staring fixedly out from a lost world. Generations of children came here: fathers, sons, grandsons. A mile or so up the coast there was the Boverton Girls Camp, with a similar layout, that also gave holidays to children from the mining areas each summer. A refurbished Boys Village was opened by the Queen in 1962, and later a conference centre was added.

The Welsh mines were shut down in the 1980s and the village closed by the Boys Club of Wales in 1990, after sixty-five years of life. Mostyn Davies, who was on the management committee from the 1960s onward, has recorded his memories. My fond hope that this was a good place is, perhaps, borne out by his insistence that the village was run for and by the holidaymakers. 'They'd have their own rules; they'd have their own government, if you like,' he says before offering his own explanation for its demise. 'It was built in the days when you found your own entertainment,' Davies tells us, but 'the era of having entertainment provided for you came along in the 1970s and '80s, and people didn't want that sort of thing.' So 'people went to Spain with their parents'; they didn't want to come to a chilly, old-fashioned 'boys' camp' any more.

I guess I'd rather have gone to Spain too. But there is more in this story than rational choices and sensible decisions. There is a more general atmosphere of loss here; a palpable sense that not just a little holiday camp but a whole way of life has been broken and disregarded. The site is tagged for demolition, and since 1971 has sat under the toxic shadow of Aberthaw coal-fired power station, which squats thuggishly on what must once have been a glorious stretch of coast. A local historian, Terry Beverton, recalls that this was 'an idyllic place of dunes, sandy tidal pools'. But that past, he says, is 'remembered by fewer and fewer', and something of value has gone. Beverton rather bitterly complains of a rising tide of indifference, much of it carried by incomers into this part of Wales; people who talk loudly about 'how Wales is such an awful place to live'.

Since its closure the Boys Village has been systematically vandalised, almost every wall covered with tags that overlay and contest each other, battling it out for ephemeral supremacy. There is a self-loathing at work here too. In large black capitals the words 'WORTHLESS' have been painted on neat brickwork; on another wall, 'I don't feel ANYTHING'. I suppose that these are just the ramblings of stoned kids, but I'm unnerved; it's as if unquiet ghosts have been disturbed. Something is being stirred up. 'Rumours have sprung up of a troubled past,' runs one oft-recycled web story about the village: 'abuse, murders in the church or fires that killed boys staying there'.

There's no support that I have found for any of these allegations, but believing the worst is, increasingly, the default condition, the safe option. It is a mood that helps turn the Boys

Village into an unwanted memory. A councillor from a neighbouring village explains that 'times have changed, and there's no need for that sort of facility any more'; adding, 'It's been derelict for a number of years, and there's antisocial behaviour there – it's attracting the wrong sorts of people.'

Paul and I have walked off in separate directions. We're lost to each other. Each step grinds down on some fallen glass or masonry, but standing still in another empty room all I can hear is my breath. And I find myself listening for him, wondering if he's close by. Not all abandoned places speak to us about our own past, but this one does; that childhood we often used to talk about, recalling the dens we'd fashion along Stacey's Lane; the long car trips to see Granny Rowe or Granny Bonnett. But there came a point when it had all been said, and the nostalgia fell away into a companionable silence. Paul and I are both in our fifties now, and it feels as if the past has been mined out, exhausted. I didn't expect middle age would be so lonely; that I'd feel so heavy inside. I set off across the parade ground, hoping to catch sight of him.

I run into Paul in a two-storey accommodation block and, at almost the same moment, stumble across a tall, black-clad young man wielding a tripod and camera. We all click into conversational mode, though the incomer has more to say than we do: 'amazing place'; 'upload onto my site'; 'I keep coming back'; 'this is one of my favourite ruins'. We don't ask him why he's so fascinated with derelict buildings because we know. It's hard to explain, but also obvious; they're interesting in a way the rest of the world is not.

He busily begins setting up his equipment and Paul and I

pick our way back into the open air, tacitly agreeing that maybe that's enough for today. We drift back into our amiable silence – the unspoken love of middle-aged brothers – and settle down for the long, winding drive up into the hills.

British Graveyard, Shimla

Across South Asia, thousands of British graveyards are overgrown and fast disappearing. There is a sadness about any neglected graveyard, but here, that emotion is especially complex: the resting places of the country's once-mighty rulers are broken and tangled over with roots, and people in Britain and India don't care. For both countries, that long, imperial moment is close by but it's uncomfortable territory, vaguely embarrassing; more than that, it has vanished emotionally.

If there is a leopard living in the graveyard, that is another reason not to go there. But that is what I was told with great certainty and a clear sense of warning about the one I'm walking down to. It is a British cemetery in Shimla, the capital of the Indian Himalayan state of Himachal Pradesh, a provincial town that, back in the days of empire, was something so much more. This lofty town, surrounded by cool pine forests, was where the British retreated to in the summer, and it was from here that the viceroys and all their numerous attendants and functionaries ruled the subcontinent.

To get to the cemetery, you drop down from the pedestrian Ridge Road, with its vistas of far mountains and ponies taking children on rides; descend precipitous stairs which zigzag down the hillside and through the bazaars; past shrines to the elephant god and holy men nudging passers-by for alms with copper bowls; past umpteen stalls piled high, many with the fat red apples that are this region's most famous crop; past a shop with great pails of sunshiny honey; then across the murderous and honking road that circles Shimla. It is, as usual, blocked with brightly painted buses and trucks. I'd been told the grave-yard is opposite the Potato Research Institute, but I can't find that and just keep going, heading for the pine forest.

This must be it. I pick my way through two broken entrance columns, and an old and broad woodland pathway opens up before me. The cemetery was inaugurated by bishops of the Calcutta Diocese in 1852, and must once have been a quiet spot, but today it is hemmed in by apartment blocks and traffic. The sun picks out thick scatters of litter. A rhesus monkey with a tiny, wide-eyed baby at her bosom dashes by, reaches safety and sits further down the path, staring at me with ferocious suspicion. I've been in this town long enough to know that I'd better grab a stick and take my glasses off. Monkeys are deified locally; a 108-foot-high orange statue of Hanuman, the monkey god, stands on a hill overlooking Shimla. But they are also treated as pests, and the previous day one had dropped on my head, stolen my glasses and dashed off up a tree. A passer-by saved the day by scattering some corn on the path. The monkey leapt down and there followed a solemn exchange; with one black paw it scooped itself a modest pile of food and with the other it handed me back my spectacles.

Without my glasses, the British graveyard takes on an even more mysterious appearance. The rubbish has faded away and I slacken my pace; after all, I've got all day. It doesn't take long before some grave markers drift hazily into view. 'Edmond Richard Purcell, Lieut 3rd Bn XII Regt Died 27 May 1871 aged 20 years 7 months. Died of injuries from falling off his horse'; 'In memory of Charles Whiteman Thomas Captain HM 21st Hussars only son of Honoratus Leigh Thomas and Sophia Boydell his wife. Of Bryn Elwy Flintshire who died at Simla 28 June 1867 in the 27th year of his age'; the dates of death range up to the first two decades of the twentieth century. What were they doing here, so far from home? Did they really imagine that this was always going to be a little bit of England? I have wandered some distance now, and it's more peaceful; a light breeze picks up the warm scent of the pine trees and the sound from the road has dimmed; and there's no chatter of monkeys. I remember the leopard, the guardian of this place, and suddenly feel a little exposed. Never mind the dead – what am I doing here?

In 2015 I learnt that my application to be a Visiting Professor at the Institute for Advanced Indian Studies, an Indian government-funded centre for research in the humanities, had been accepted. I would be staying at the Institute's palatial gothic home in Shimla, the once-named Viceregal Lodge from which this nation was governed from 1886 to 1946. With visions of tea, sunlit verandas and witty, post-colonial badinage I flew into an airport that, on the map, looks nearby. Many hours later, it was pitch black and I was still in my taxi, bouncing and veering up ever-smaller roads. Eventually we stopped; an elderly man loomed by the window with a huge bunch of keys and got in beside me. We drove off again to what, in the darkness, appeared disconcertingly like a hut in the middle of a forest. The driver and the older gentleman bade me farewell. I ate some of the cheese crackers I'd packed in my luggage and, very relieved to have finally arrived, fell asleep.

In the morning sunlight, things were much clearer. I was in a hut in the middle of a forest. It was a long hut, divided into three units; I was in the middle one and the other two were empty. I was utterly alone. In the little kitchen was a gas hob: taking the hint, I went down through the trees to find something to cook. The track soon took me to a narrow road, each side of which was crowded with little booths selling cooking pots, toiletries, lots of crisps and various heaps of fresh fruit and vegetables. Pleased that I was not as isolated as I'd thought, I was soon lugging back my bounty: a bag of eggs, some rice and apples, a jar of pickle and plenty of crisps. But I was puzzled. Was this to be my life for the next month? I could now see that the hut was a rather decayed example of early twentieth-century staff accommodation, and that the whole hillside was dotted with similar cabins.

Climbing upward behind my new home, a path led to the rose gardens and manicured lawns of the sprawling Viceregal Lodge (it is still known by that name, though it has been officially retitled Rashtrapati Niwas, or President's House). I came to know it well: its panelled rooms and lofty libraries with leaking roofs; endless empty corridors; occasionally an airy office in which sat a man filling out variously coloured forms. On the walls were conspicuous gaps where the most egregious symbols and regalia of British occupation would have once hung. But nothing much had taken their place, and the atmosphere was strange and heavy.

Today the Viceregal Lodge is a disconcerting building, alive with too many ghosts. I felt, maybe, that I was one of them; or, at least, a possibly unwelcome reminder of an overbearing past. The Fellows of the Institute are largely young postdoctoral scholars of a friendly and earnest disposition who live together in a lovely rambling house with roses, tea and a veranda. The seminars I diligently attended were on Indian history, ethnology and politics, and took place round a huge oval table. They were impressive; the speakers impassioned, self-critical and original. Yet after each event everyone disappeared, and I was left wandering down to my dark dwelling like some friendless spectre returning to its tomb.

My patience finally snapped when I pulled on my clothes one morning to find that large, if docile black insects had begun camping inside them. I dragged my suitcases along the Ridge Road until an ancient and tiny porter grabbed hold of them, put them in a shabby baby buggy and, with the sweat leaking out of him, hoisted them up the steep slope to a hotel at the other end of town. I paid off my guilt with a huge tip and wallowed on my lovely bed: everything was now good. I felt I'd shed my responsibilities to the Institute. I promised myself that I'd eat out every night and spend my days happily roaming around Shimla.

And I did. Almost every corner of Shimla reveals a new vista of half-timbered or stone gothic Victoriana. Many of the buildings are in government hands and are well maintained, but dozens are caving in; their windows and wrought iron gates hanging off and grass and trees pushing their way into every opening. Yet it would be a mistake to picture Shimla as an architectural Ozymandias, a shattered visage bearing the cold sneer of long-gone command. For the British presence here is a living thing; it has been absorbed into the town's culture, been taken ownership of. You can see and hear it in the noisy parties of smartly uniformed schoolchildren chatting in English, who, several times a day, pour out onto the streets. Or in the formal manners of so many of their parents; the handshakes and cricket club-style bonhomie. The town is awash with uniformed functionaries of all sorts, from the policemen blowing their whistles to the soldiers strutting about at checkpoints and the railway staff who welcome in the famous steam train that puffs its way up to Shimla from the hot plains to the south. It's a safe, charming and multifaceted town, and I never felt unwelcome anywhere I went, even when wandering very much off track in pursuit of the last resting places of some of those British people who thought of Shimla as home.

The British Association for Cemeteries in South Asia tries to document the many disappearing graveyards of the subcontinent, but it is a mammoth endeavour. The burial ground with the leopard is called Kanlog Cemetery, but there are many others spread across the district: just this one town has five

British burial sites. They have met different fates. The Nun's Graveyard, with its large cenotaph to Colonel Parker of the Bengal Army, lies in a wood of cedar trees behind a convent school and remains in good condition. The others are derelict, with most of the graves lying shattered – those that have not already disappeared. Some are just traces; not even that: they are long gone, have been built on and forgotten. There used to be a British cemetery behind the noisy central bus station, where a hotel now stands. A town guide reports that, 'The cemetery was very neat and well-kept by a caretaker till the beginning of the 1960s. Plenty of violet flowers bloomed during spring.' But it also tells us that 'antisocial elements and some homeless people started encroaching the land and indulging in vandalism by stealing the wrought iron decorations and marble headstones'. Finally the graveyard was occupied by Tibetan refugees and became 'a shanty town of Tibetans. No sign of it remains now and no one remembers it.'

My time in Shimla has left me uncertain about these endless avenues of British graves. Initially I felt rather shocked that they were being left to crumble; that any cemetery could be so disregarded. Yet there are more pressing matters in India than graveyard conservation, and the country is layered with multiple claims on the past. They can't all be preserved. My attitude has cooled; I'm less sure. Perhaps these decaying kingdoms are, after all, best left to the ghosts and the leopard.

Dau Movie Set

I began writing about the world's strangest movie set years ago, but gave up. Much like those who have been waiting, for over a decade now, for the set's offspring, a film called *Dau*, to come out, it was frustrating and confusing. More than that, there is something queasy and troubling at work here. *Dau* is often called the real-life *Synecdoche, New York* (2008), a reference to Charlie Kaufman's brilliantly unsettling comedy about a fictional film director, Caden Cotard, who builds a small version of New York and turns his actors into its real-life inhabitants. Cotard is stern but inspirational: 'I won't settle for anything less than the brutal truth. Brutal. Brutal,' he barks at his cast.

Kaufman says something very telling about his main character: 'The only way he can reflect reality in his mind is by imitating it full-size.' This is precisely what the young Russian film director Ilya Khrzhanovsky achieved with *Dau*. Between 2006 and 2011, replete with a windfall of large grants, Khrzhanovsky recreated Moscow in an abandoned sports complex in the Ukrainian city of Kharkov. The film was about the unconventional life of a Soviet physicist called Lev Landau ('Dau' was his nickname). He was adamant that his cast and staff must fully inhabit the drab routines of 1950s Moscow. They had to live on site, and immerse themselves in the period and the place. The set wasn't merely a backdrop but a functioning place. Each 'actor' carried on with his or her character's role whether on camera or off. It was a place of food shortages and ill-fitting clothes. Women were asked to wear Soviet-model cloth tampons. Stepping out of character incurred financial penalties. The mention of modern communications technologies, such as the internet and CGI, carried heavy fines.

In this brave new world, loudspeakers blasted Soviet-era propaganda, and there was a newspaper producing Soviet news, a Soviet hairdresser, a Soviet cafeteria and a laboratory where real scientists performed real experiments on real animals. Fourteen children were conceived on 'set'. But it wasn't a set any more. When Khrzhanovsky wanted to stir his listless and increasingly hopeless cast, he would have one 'actor' report on another, and what appeared to be the 'KGB' would come and make arrests. It wasn't the KGB, of course, but nor was it a group of actors, pretending. It appears that a home-grown secret police emerged organically, ready to denounce their neighbours. One 'real' scientist who had been invited to live on set told the Russian newspaper *Kommersant* about his own late-night visit:

When they came into our home I still didn't know who they were coming for. I lit a cigarette, although I hadn't smoked for 15 years. I got up, half-dressed and prepared myself, then waited while they climbed the stairs. When I came out onto the landing and it turned out they hadn't come for me, but for the neighbour, I looked like someone who'd just died. My face was white . . . I was literally in a bad way, it was terrible, I experienced emotions on a real scale.

Those who were 'arrested' had to share cells with criminals hired by the director from the local jail.

There have been plenty of spectacular movie sets built before. The film industry relishes tales of directorial excess and insanity. They are usually exaggerated, but *Dau* is the real thing. Khrzhanovsky did not just blur the border between representation and reality; he tore it to shreds.

Today *Dau* has passed into legend. Even if it is never released, it has already secured its place as the defining 'runaway' film production of all time. But that is too small a word: *Dau* is not just another example of indulgent budget overshoot. We can find out more from one of the few journalists who got onto the set. In 2011, not long before the set was ordered to be destroyed by the director (he got a load of Russian neo-Nazis to invade and wreck it), the journalist Michael Idov was given a rare tour by the director. The formalities started at once: 'A silent guard observes my typewritten pass bearing the Soviet hammer and sickle and date-stamped April 28, 1952. Another frisks Khrzhanovsky, without betraying any deference or even recognition.' Idov finally makes it through to the set: 'Before me is an entire city, built to scale, open to the elements, and – at 1 a.m. and with no camera in sight – fully populated.' Idov notes that 'Two guards walk the perimeter, gravel crunching under their boots. Down the fake street, a female janitor in a vintage head scarf sweeps a porch.'

The set was 'roughly the size of two football fields' but built on different levels, allowing a sense of the whole city: 'A coliseum-like stadium looms over two drab residential buildings. Atonal cello music squalls across the city, issuing from pole-mounted loudspeakers.' In normal film sets the doors are fake; they open onto plywood walls. Not here. Here they opened onto corridors which led off into rooms in which were found beds and tables and cabinets. And in those cabinets were drawers, and in those drawers were things. It wasn't a facade. Was it even a set? At one point during Idov's tour, Khrzhanovsky opens the front door of one of the residential buildings and it becomes clear that the 'guts of the set are as elaborate as the set itself. There are hallways that lead to apartments, and in the apartments there are kitchens, and in the iceboxes food, fresh and perfectly edible but with 1952 expiration dates.' This wasn't just attention to detail; it was a full-scale reanimation. Obviously delighted to show off his world, Khrzhanovsky 'opens cupboards, drawers, closets, showing me matchboxes, candles, loofahs, books, salami, handkerchiefs, soap bars, cotton balls, condensed milk, pâté'. His finale is flushing several toilets. 'The toilet pipe is custom width,' he tells his visitor, 'because it makes a difference in the volume and the tenor of the flushing sound.'

Why would any director need to go to such lengths? Khrzhanovsky explains that all this artifice allowed him to 'elicit the needed emotions from his cast in controlled conditions, twenty-four hours a day'. But this is nonsense. No film-maker needs twenty-four-hour emotions. *Dau* wasn't about what 'needed' to be done; it was barely even about making a film. From early on, rumours that the set was deeply unusual soon started circulating in the (real) city and in movie-making circles. A figure of 210,000 cast and crew members was bandied about, although this turns out to be merely the number of vetted candidates. Moreover, the inflexible rules that governed the set were not just strict; they went far beyond anything either reasonable or defensible. One Russian movie blog reported that Khrzhanovsky was 'A prolific womaniser' who 'interviews potential female cast members with questions like "Can you come up to a guy in a club and sleep with him without finding out as much as his name?" or "Are any of your friends whores?"' 'Working here,' explained one Khrzhanovsky staffer, 'is like being that guy who wanted to be killed and eaten, and finding a maniac who wants to kill and eat you. Perfect reciprocity.'

The *Dau* set was designed to grind down the differences between acting and not acting, and between a staged place and a real place. It is that transgression that I think makes it so appalling and so wonderful, and why its story will continue to be told as long as people talk about films. In 2015 Russia's Ministry of Culture, who had part-funded *Dau*, had had enough and called for the return of their investment of US$340,000 (almost £265,000) plus US$120,000 (almost £93,500) in interest. However, other European backers remained stubbornly faithful. In *Synecdoche, New York*, the director had a similarly bottomless chequebook. At the start of the film he wins a vast award from a charitable foundation to spend whatever he wants in the name of his art. So he too builds a city within a city. It seems that too much money makes too much possible, and unhooks the bond between representation and reality.

One London-based post-production editor who worked on *Dau* is starkly damning: 'I can confirm that this project is indeed insane, and not entirely in a good way.' Yet the same editor also admits to being 'fascinated by this project'. It has a grip, it forces admiration, while creating deep unease. All I've seen is the trailer, and it's already clear that the ultimate irony is that the film *Dau*, the supposed end product of all this obsession with authenticity, doesn't even rely on realism. It employs a lot of close-ups of grim, sad faces; lots of intimate facial expressions showing deep resignation. A press briefing from Khrzhanovsky tells us that 'this story could take place here and now, at the beginning of the 21st century'. What he means by this is that *Dau* is 'a story about freedom, and mostly about internal, personal freedom; about secret desires that a man permits or doesn't permit himself to realise; about how much a man can allow himself and what he has to pay for this'. Of course, it's a statement that tells us far more about Khrzhanovsky than it does about Lev Landau. Khrzhanovsky's 'secret desires' and what he will 'allow himself' are now well known and infamous. This, he says, is a film 'about how to live with a Genius, how to love a Genius, how to tolerate a Genius'.

Magical London

Magical geography is the oldest and most resilient form of spirituality. For thousands of years people have looked around them and seen the hills, rivers and woods not as inert things but as seething with restless spirits and uncanny powers. The conviction that we inhabit a living, mysterious landscape is deep-rooted. It's an intimate, highly local connection that was almost severed with the development of the religions we are now familiar with. Conventional religions try to lift up our sights from the earth and towards universal truths and salvation in another realm. Magical geography never had much interest in salvation, or in alternative kingdoms such as heaven and hell. It was a place-bound and place-loving kind of spirituality, and for many millennia it met the needs of a place-bound and place-loving species.

Our curious new century is birthing a new generation of geomancers, psychogeographers and landscape mystics. Increasingly they are not content to restrict themselves to safely residual heritage locations, such as stone circles on misty plains. They are intent on finding enchantment amid the built-up and the built-over. I must admit I warm to them; these eccentric communities of urban ley line hunters, roadside shamans, soothsayers and spell-makers are like spring returning. I don't believe in the literal truth of their mystic claims, but the undogmatic, nature-shaped character of their wandering visions is somehow irresistible.

London has emerged as the navel – or, as the new mystics like to say, omphalos – of this new urban magic. The best-known example is the sacred lines of power, or ley lines, that have been mapped in many different ways across the capital, each attempting to identify a counter-cartography that connects sites of natural and spiritual energy to real-world systems of power. Freddy Silva, a researcher of 'alternative history', explains that 'around and beneath the present convoluted metropolis of buildings and roads there is still evidence of ancient temples and sacred sites, each purposefully placed and aligned, and taking advantage of universal laws associated with the manipulation of subtle energy'.

The ley line maps of David Furlong, another researcher of earth mysteries who began his own 'esoteric training' back in 1968, inscribe 'primary' and 'secondary' triangles across the city. To get a better sense of what he's up to, let's look at his 'secondary triangle'. It is an equilateral triangle with its apex at 'Boudicca's Mound, Hampstead Heath', and a base that 'links the Tower, Southwark Cathedral, the Palace of Westminster, Westminster Abbey and the West Brompton Cemetery'. Furlong adds another claim: this triangle, he says, 'was understood and appreciated, at least until the eighteenth century', a fact 'evidenced in the alignment of John Nash's Avenue in Regent's Park, which is very accurately aligned to the Boudicca's Mound and also exactly bisects the triangle'. The idea that early modern architects and planners incorporated magic into their designs also appeals to Freddy Silva, who tells us that the city's great seventeenth-century planner and builder, Sir Christopher Wren, 'rebuilt the city using a blueprint based on the ancient system of sacred geometry and Kaballah'.

The ceremonies and 'traditions' of contemporary magical geography are not really that old. The idea of 'ley lines' was the invention of Alfred Watkins, an amateur archaeologist, in a book which came out in 1925 called *The Old Straight Track*. One of modernity's more interesting consequences is the creation – some might say fabrication – of a more primitive and authentic past. But this insight doesn't diminish the importance, or necessity, of these new forms of 'earth mystery'. We're not being offered dogma, but ideas that imaginatively reclaim and disorientate the ordinary landscape, making it more curious and multidimensional.

Over the last couple of years I got so obsessed by this fashion for occulting geographies that I turned it into an academic paper which, to my surprise, got accepted in the *Transactions of the Institute of British Geographers*. Far from regarding the topic as outlandish, the journal's peer reviewers seemed a little fatigued by it. One referee started their report with, 'Like many, I am weary of the term psychogeography.' But they had to admit that I'd bumped into a phenomenon of consequence, if only because so many books, websites and performances have appeared over the last ten years that have striven to re-enchant London. The doyen of this scene is Iain Sinclair, who has mined the arcane byways and lore of London streets for decades. But there is a newer generation of London writers fermenting their own geo-alchemy. One of my favourites is Nick Papadimitriou. He's been called London's post-post-modern shaman and his book *Scarp*, which appeared in 2012, is unembarrassedly prophetic.

> Slip, Motorway, round my ankles if you must; drag me into your petroleum future. You will pass too, ending crotcheted by red leaves of herb Robert, stars of cow thistle. I see your car crashes. I see economies collapse. I sense the unspoken family secrets; I see the white cow-gate lit by sunshine. I am the centre.

For Papadimitriou, such moments of enchantment are a necessary act of imaginative possession in a landscape that does not, ostensibly, want him. 'How did we end up with a city where walking is so hard, where the land is so hostile?' he asks, before recounting one of the many humiliations that have befallen him. Papadimitriou is arrested, goaded in police cells and courtrooms and, unlike Sinclair's clubbable forays, appears friendless. His walks are littered with references to failed attempts, to 'the fear of this barren flinty earth' and to feelings of being 'bewildered'. It is in this context that he summons magic as a vital resource, not just for him but for the land that he values.

In the wake of *Scarp*, other magical reclamations of London began to appear, such as John Rogers' *This Other London*. At first it appears a bit lightweight next to Papadimitriou's urgent prophecy. But one of the endearing traits of London's new magical geographers is their weakness for comedy. Rogers' ten walks of 10 miles each aimed 'to cover as much of the *terra incognita* on the map as possible'. He's particularly drawn to the disjunction between the ancient and the modern, the magical and the prosaic, and to bathos. 'I was attempting to access the woodland spirits and reawaken my pagan instincts,' he reports, 'when I found myself back on the A206 Woolwich to Erith Road buzzing with afternoon traffic.'

Another wonderful example of London's new mystic crop is Gareth Rees' *Marshland: Dreams and Nightmares on the Edge of London*. 'London is dreaming again,' says Rees. He argues that extraordinary forces have been released because the city is psychically disintegrating. Rees wants us to know that London is a place of 'entropy' where 'weirdness' is breaking through the surface of normal life. Rees is another walking writer, creating and documenting footsore rites and hallucinogenic ceremonies. The site he communes with is in the East End of the city: the marshlands of the Lea Valley and its fast-disappearing traces of industry. Rees' London is a city of vengeful ghosts who are plaguing us with bad dreams: long-dead refugees and factory workers rising from the mud and silt; zombie protesters against environmental despoliation and gentrification. Rees' 'marshland' is 'a hole in London . . . strewn with ruins, rubble and wild flowers. It swallows up the city's time.'

The disused Middlesex filter beds, zoned for demolition, are a particular site of fascination for Rees. 'Inside its gate was a network of stone ramparts, split and crumbling,' he writes, adding, 'In the centre was a giant stone circle, like a sacrificial altar.' Such abandoned places accrete both psychic and material debris, which Rees deciphers as forms of restless energy. Surrounded by dimly lit and unclear graffiti – 'a warning sign? An occult symbol?' – he knows that he is 'not alone'. To capture this landscape Rees adopts a geo-ceremonial poetic structure, in which the names of objects are spread on the page as they are spread on the ground. It ends up looking like a chalked spell or thrown runes. 'Sheltered at the pylon's foot', he records the following:

Coke bottles   polystyrene plates

                                          plastic bags

                    The Mirror

a sock, soiled   fragments of cardboard box

                  a whittled stick

What are we to make of all these attempts to re-enchant London? Some might read them as the confused spiritual urges of a secular city; a reflection of the chaotic delta of beliefs that occurs in the absence of 'organised religion'. But that would be to misread the variety and insistent individuality, the sheer quirkiness, of the new magical geographers. Far from being confused, they strike me as thoughtful, open-minded, playful and curious about the world; all attributes shamefully rare in 'organised religion'.

The desire to recognise the landscape as a living, mysterious presence is as old as the human story. It's still a vital force in many cultures: East Asia has an unbroken and ancient tradition of geomancy, and many indigenous peoples maintain that their bond with the earth is their most precious possession. In the Western world, in London, we are trying to reinvent this connection. The results are, predictably, something of a ragbag. But these walkers and wonderers are on to something; they are pushing at an open door. There's a feeling out there, in the air; a need for a more generous, more ineffable and enchanting relationship with the city.

Tsunami Stones and
Nuclear Markers

The stern stones, chiselled deep with Japanese inscriptions, stare out to sea. For up to six hundred years they have been grimly repeating an urgent message, often in the form of a simple warning: 'Remember the calamity of the great tsunamis. Do not build any homes below this point'. There are hundreds of these tall stones in Japan, and below nearly all of them sprawl streets, roads and houses, waiting for the next wave and to be reminded of the stones' wisdom, yet again.

The stones can be found on the north-east of Honshu, Japan's main island; the same shore that on 11 March 2011 suffered the worst casualties from an earthquake and tsunami that killed 15,894 and left another 2,562 still missing. The Japanese are used to the power of nature, but the shock of this disaster, which is often referred to by the month and year as 3/11, has been profound. One of the consequences has been a renewed interest in the stones and an anguished search for new ways to send a message to subsequent generations.

How can we warn future generations? What can we do to create a mark in the landscape, some permanent alteration, which will make them sit up and listen? This is not just a dilemma for countries that suffer from tidal waves. The most profound challenge is for those many countries looking for somewhere to put nuclear waste, which will remain deadly for a hundred thousand years or even longer. How do we tell people to keep clear, to stay away from radioactive disposal sites, across such deep reaches of time? Recently the nuclear industry has been coming up with possible answers to this question, in part by drawing on the story of the tsunami stones.

The tsunami stones range from three to ten feet tall, and they have not all been ignored. In the small town of Aneyoshi people say that it was the stone that saved them. It was erected after a tsunami in 1896, and the villagers heeded its warning not to build beyond it. They listened because the stones are not merely information boards; they also have a memorial and spiritual function. The oldest contain Buddhist teachings which link natural calamities with karmic retribution. Later ones show a Shinto influence. About 120 years ago the stones' messages began to take on an earnestly practical tone; such as, 'if there is an earthquake, think only of yourself and run to high ground'. After 3/11 the National General Association for Stone Shops in Japan erected 500 new coastal stone monuments, similar in look to the old tsunami stones but with an English translation and QR (Quick Response) codes that link to information about the 2011 disaster.

These new stones, like the old ones, are complex objects, combining the functions of warning, memorial and shrine. Going by past experience, many worry that they too will be ignored. The aftermath of 3/11 saw a popular movement to maintain more tangible and shocking reminders by preserving some of the over 1.1 million buildings that collapsed or were damaged on that day. Proponents of these 'warning ruins' look to the most famous example of the type, Hiroshima's Atomic Bomb Dome. Originally an exhibition hall, it was the only structure left standing near the epicentre of the nuclear explosion that destroyed the city. The surviving shell is preserved as an exhortation for peace and as a commemorative monument.

As impressive as such ruins may appear, they will not last long; at least, not when compared to the many generations to come that need to be warned. This is especially pertinent when we are trying to warn people in the future about the dangers posed by the unwanted by-products of the nuclear industry. The disaster at the Fukushima nuclear power plant has made towns across Japan very wary of wanting anything to do with the 16,700 tons of radioactive waste that the country has in 'temporary storage'. Someday it will have to be buried, and that spot will be a danger zone for what, in human terms, is forever. The country's nuclear authority has been cajoling communities to be patriotic and offer up suitable land: 'We should feel grateful for the community that's doing something for the benefit of the whole country,' says nuclear official Takao Kinoshita, adding rather ominously, 'and respect their bravery.'

The international scientific consensus is that nuclear waste
has to go deep underground, placed into geological layers that
are stable and sealed from seepage. The world has 437 operating nuclear reactors, but only two countries, Finland and
Sweden, have managed to reach a public agreement on where
to dispose of the waste. To which future generations might say,
so what? After all, it is they, not us, who will be dealing with
this legacy; humans living up to and beyond four thousand
generations in the future, people who will be as distant from us
as we are from the early hominids.

Examples of nuclear warning stones do exist, but, given the
immense challenge, they are meagre things. In a public park
outside Chicago stands a warning stone with the message:
'Buried in this area is radioactive material from nuclear
research conducted here'. Its main function is communicated in its last line: 'There is no danger to visitors'. One
wonders what 'visitors' in the year 3000, or 20,000, will make
of this. Nothing at all, of course; because in a few hundred
years the stone will be illegible and probably long gone.
Interviewed by the *Financial Times*, Patrick Charton, head of
a 'memory division' at the French nuclear agency, asks, 'How
do you write a message that lasts thousands of years? What
language do you use? What do you even say?' Answers have
come in various forms, from the loud and monumental to
the unnervingly quiet. In the latter category is the Finnish
response. Their deep geological repository is at the top of the
Gulf of Bothnia and, once complete and full, it is designed to
disappear into the landscape. It is believed that, because this
is a remote and frozen place, no one will ever disturb it, and

that any kind of marker would attract unwanted attention. Kai Hämäläinen of the Finnish Radiation and Nuclear Safety Authority explains: 'the idea is that the facility will be safe forever, even if the memory is lost'.

Try as I might, I can't grasp this argument. It is based on too many assumptions to be any kind of solution. More plausible responses have been developed by the US Department of Energy, who, back in the 1990s, asked a range of futurologists to design a permanent marker for a waste site in New Mexico. The result, which remains as yet just a plan, is anything but quiet. It is a red-faced bellow into the future; a 'message wall' inscribed with a series of fearmongering messages in seven languages, including Chinese and Navajo. The English version reads like an incantatory and disturbing poem:

> This place is a message, and part of a system of messages,
>   pay attention to it!
> Sending this message was important to us.
> We considered ourselves to be a powerful culture.
> This place is not a place of honor.
> No high esteemed deed is commemorated here.
> Nothing valued is here.
> What is here is dangerous and repulsive to us.
> This message is a warning about danger.
> The danger is in a particular location.
> It increases toward a center.
> The center of danger is here, of a particular size and shape,
>   below us.
> The danger is still present in your time, as it was in ours.

The danger is to the body, and it can kill.

The form of the danger is an emanation of energy.

The danger is unleashed only if you substantially disturb this
   place physically.

This place is best shunned and left uninhabited.

Since we have no idea if any of these words may be read or understood in a hundred thousand years' time, the plans also include a range of symbolic landscapes, each designed to communicate horror and danger. The look of them can be guessed by their names: Landscape of Thorns, Spike Field, Spikes Bursting through Grid, Leaning Stone Spikes, Menacing Earthworks and Forbidding Blocks.

A few years ago a 'Preservation of Records, Knowledge and Memory across Generations' group of experts was created by the Nuclear Energy Agency in Paris. They have conducted a study on the Japanese tsunami stones and concluded that they didn't work. Having failed to agree on any type of marker that would work better, the new think tank has started to put its faith in more intangible forms of warning. They have turned to artists to come up with ideas. To date, these include creating children's songs about radioactive waste that get carried down generations, and building a creative 'laboratory' above the waste sites in which each generation can think up new ways to explain the problem of nuclear waste. Another idea is to breed 'Ray Cats', genetically engineered felines that will start glowing when they are near radiation.

There is a disconcertingly playful quality to these 'solutions'. It seems that the challenge is so big, the experts have given up and are now entertaining themselves. Rather than publish details of nuclear nursery rhymes and light-up cats, these 'memory workers' might as well put down their pens and issue a simple statement: 'It can't be done.'

The tsunami stones, old and new, do important work, but it is time to admit that we can't warn the distant future; we can't buy off our guilt with stones, spikes, songs, luminous cats or anything else. With that acknowledgement should come another: the choice is between saying 'let the future sort it out because we can't', or 'we shouldn't be creating lethal hazards for unborn generations'. The former position is so obviously irresponsible that it seems to me we are stuck with the latter. Our efforts need to go into creating cleaner types of energy supply. The tsunami stones determinedly exhort us to learn and listen; and one of the things they are saying is that we need to take responsibility for our own actions, and not pass the buck to the future.

V.
Hidden Places

The age of Google Earth fools us into thinking that everything is seen; that the dark corners of the world have gone. In fact, all that surveillance from above only makes the shadows deeper. They are certainly deep in Cairo's Garbage City, where we find a hidden community sifting and recycling rubbish, offering a service which is as vital as it is disregarded. Any user of Street View should know about its no-go areas, places which form an alliance of invisibility between the very rich (Hidden Hills) and the very poor (the slums of Wanathamulla). The secrets of the map are again to the fore when we explore Trap Streets, those deliberate mistakes that help map-makers spot plagiarists, and the Uncharted Congo, an anachronistically blank space at the heart of Africa. The close association between wealth and clandestine places resurfaces at Flat 2, 18 Royston Mains Street, Edinburgh, where hundreds of 'shell companies' and other tax avoiders feed from and into the UK's archipelago of tax havens. A different, if equally antisocial example of an under-documented and overlooked zone is that of the anti-pedestrian cobbles and jagged paving that form the Spikescape. These aggressive microtopographies try to guide our movement in the city. No discussion of hidden places would be complete without a nod to its most famous and aggressive form, the secret military base. My example comes from China: the Yulin Underground Naval Base. Digging down also bring us Under Jerusalem, although this story is as much an excavation of my old diaries from the time – thirty years ago – I visited that fabled but fractious city. We conclude our adventures with three watery examples of hidden places. The first is under the North Sea and is called Doggerland. We then head to the New Arctic, where new rivers, canyons and shipping routes are all opening up. Finally we dip down in much warmer waters to find Jacques Cousteau's old lair, Conshelf Undersea Station.

Garbage City, Cairo

A giggling girl of eight or so sits amid a nest of crunchy
plastic, picking out the crisp packets. I've arrived in
Garbage City on a cool late-March morning, the car bumping
up a broken roadway, dodging boulder-sized bags of rubbish
and yawning dogs. Once on foot, I soon find that the streets
of Mokattam Village, Cairo's Garbage City, are a visceral and
pungent experience. Men, women and children pick through
various types of rubbish, searching out and sorting the recy-
clable material.

I nip sharply out of harm's way as a hidden figure cutting
up car panels sends golden sparks flying out from a doorway;
nearby, young men are perched on sacks crammed with
multicoloured rubber pipes, tucking back the spilling, intes-
tinal tubes. In the darkness of other ground-floor workshops,
conveyor belts and shredding machines are churning and

grinding, gulping down dirty cloth or plastic. Although it's not yet hot, flies hang in black clots, matting the piles of garbage and giving the momentary illusion that the trash is sorting and shifting itself without the aid of any human hand.

Here live the Zabaleen, Arabic for 'garbage pickers'. White crucifixes are painted on many walls, statues of Mary and Jesus fill niches otherwise occupied by rubbish. Mokattam Village, sitting close to central Cairo but on a craggy hillside in the shadow of Mokattam Mountain, is a Coptic enclave, a unique Egyptian branch of the faith that is almost as old as Christianity itself.

A constant stream of fresh trash is arriving in the village; it is collected, starting at dawn, from the streets of Africa's biggest city, which sprawls below. It is carried in on wooden carts pulled by donkeys as well as in small battered trucks. Like some morsel dragged into an ants' nest, as soon as a bag arrives it is busied over, dissected; some bags are carried up by chains to higher storeys; others are left for sorting on the ground.

I've been sticking to the main streets, but my amiable guide, Tarek, who looks younger than his thirty-two years, decides that my interest in what he calls 'one of Egypt's many hidden places' needs to be more fully sated. We head, a little uncertainly, up a side alley. The rubbish is now coming in at us from all sides; it's like entering a cave of garbage, and the atmosphere turns darker. Who are we to wander in here, where outsiders rarely come? And that includes other Cairenes; they too never come here. It's an open secret and, for some, a source of shame in the city's midst. Reactions to our intrusion, already wary, are turning defensive. 'They are very, very touchy,' says Tarek after his amiable greeting is rebuffed by two youths, and a woman shouts, 'No photographs, don't let them take photographs,' out into the street.

We are both trying to look like we know where we're going, but this is a labyrinth, each lane disappearing off into vertiginous bales of trash. After some hasty doubling back, Tarek calls for our driver to come and collect us. But waiting for the car, my eyes and nose settle down and I begin to notice other things. Standing at the crossroads is a cart piled not with rubbish but with a great pyramid of strawberries, and other barrows ferry fat garlic bulbs; there are numerous tiny cafés and stores; trays of puffy Egyptian flatbreads are being carried to and fro. And the people, far from being dressed in rags, look just as well shod and groomed as other Cairenes. Young women and girls stroll by, laughing and chatting, heads uncovered, wearing bright, pristine shawls, the older women in long, black embroidered gowns. There is so much more here than misery and victimhood.

I'd already seen and learnt about the formidable nature of the Zabaleen earlier that morning. I'd started my visit high above the village at the rock church and amphitheatre of the Saint Samaan the Tanner Monastery. With sweeping views over the hazy, car-honking city, it is claimed to be the largest church in the Arab world. The open-air cavern, one of several on the hillside, began to be dug out in 1974. It is a holy site, holding the saint's bones in glass casks and, even more miraculously, a ceiling carving of the Virgin Mary that was said to have been discovered in situ while the cave was being excavated. The cliff faces are also decorated with numerous reliefs of biblical scenes by Mario, a Polish man who has worked in the community for many years.

As we arrive we can hear liturgical chanting. It is the 'forty days'; a funeral service that, according to Coptic tradition, is held forty days after death. Descending the long entrance ramp, we pass a line of about two hundred men, each intoning responses to the priest's incantation, all in the Coptic tongue. Squeezing up to the back of the amphitheatre, I can now see that this line is formed by the male relatives of the deceased and, this being the end of the service, the other mourners are all waiting to file past, shaking hands and consoling each relative as they go. The men wear a variety of sombre smocks and shawls, indicating their families' place of origin in Egypt. For the Zabaleen are migrants, arriving in Cairo in the mid-twentieth century, then being pushed out to this arid ground in 1969. Below the men are the women, queuing quietly in the same fashion, all clad in black, and beyond them I can see the altar, set on a wide stage, around which are huge photographs of the dead: an elderly couple and a younger man. I never do find out how they died.

After the mourners have filed away I go down to talk to Maged, an ever-smiling Zabaleen custodian of the church, the back of his blue windcheater emblazoned with 'ASK ME'. Maged's chatter is full of the message of God's love, and he reels off a list of miracles – from moving mountains to holy visions – gleeful and excited at the proofs of divine power and the prospect of salvation that he sees all around him. The existence of the church is itself a kind of miracle, another source of hope. 'This is the church of the Zabaleen!' Maged boasts; 'This

is ours', although he pays effusive homage to all those who
have helped, especially Mario and the Belgian-born nun, Sister
Emmanuelle, who worked to bring a hospital, schools, water
and electricity to the community and whose memory brings
tears to his eyes. She died in 2008 a few days before her one
hundredth birthday. Before these 'blessings', Maged says, 'the
water came up by donkey, it took five hours'. Now he smiles:
'We have a good life here.'

Anyone who has trodden the lanes of the village below
might wonder at that judgement, but over the last two decades
the lives of the Zabaleen have been transformed and their job
of work has gained some recognition. It is estimated that they
handle one-third to a half of the thousands of tons of garbage
the city creates each day, and they recycle 85 per cent of what
they find. This is a far higher recycling rate than is found in
most Western cities. Wael Salah Fahmi, Professor of Urban
Design at Helwan University in Cairo, argues that the Zabaleen
'have created what is arguably one of the world's most efficient resource-recovery and waste-recycling systems'. Within
the international development community, the Zabaleen are
famous. They are 'far ahead of any modern "Green" initiatives',
according to *Garbage Dreams*, a recent documentary which
follows the tribulations of three teenage Zabaleen boys.

Yet these plaudits have not shifted perceptions among
the Cairenes. The public view of Garbage City is tinged with
disdain, not just because of the smell but owing to the Zabaleen
tradition of using pigs to consume organic waste. Egypt is, for
the most part, a more tolerant place than many of its neighbours, but it is a 90 per cent Muslim country and attitudes to

the Zabaleen can quickly veer from wrinkled noses and mild amusement to outright contempt. A government-imposed mass slaughter of the community's pigs in 2009 was justified in terms of containing swine flu but, given there were no cases of swine flu in Egypt at the time, it was understandably interpreted within Mukattam Village and the wider Christian community as having been driven by religious hostility.

The squeeze on the Zabaleen went further a few years later when the government arranged garbage collection contracts with several international companies. It soon transpired that the eager contractors had not understood that garbage collection is a complex and laborious job, and they pulled out. Cairo had to turn back to the people it had just turned its back on. It was at this crisis point that a potentially great idea was born, that of professionalising the Zabaleen by getting them to set up small businesses which could win contracts with the local authorities. Soon hundreds of these enterprises had been created.

Yet on the day of my trip, I learnt that this scheme had itself run into trouble. 'The Zabaleen are very angry,' Maged told me. 'Now we have a big problem. Ten days ago there was a government decree that means everyone can take our plastic, our cardboard and metal. Now we don't find recycling; we find only organic food waste.' What had happened is that a new payment scheme for recycling was about to be trialled, a scheme open to anyone and everyone. Kiosks would be set up around the city which would buy up waste, bypassing the Zabaleen.

We drive away down the mountain and into Cairo's melting tangle of cars and noise. The streets are full of rubbish; people chuck it from car windows, out of their pockets, sweep it out of their houses. It's a filthy city. Great rafts of trash float in the Nile and shoal the sands around the Pyramids of Giza. There it stays, just like it does in thousands of other cities. I know when I get home to England I'll find my own street, my own neighbourhood strewn with rubbish, just like it always is. This is not a problem unique to Cairo. What is unique is that this city has a community dedicated to clearing up other people's mess and making use of it. It seems obvious that they should be regarded as one of the city's greatest assets, not a source of shame or pinched noses but of the proud boast: 'Cairo, city of the Zabaleen'.

Off Street View: Hidden Hills and the Slums of Wanathamulla

The very rich and the very poor have one thing in common: they're not on Google Street View. Picking up the little orange humanoid on Google Earth and tumbling it down to Street View level is something I do all the time but I've been noticing that, while Street View keeps extending its empire, there are stubborn gaps in its coverage.

Hidden Hills in California is one example. Estate agent websites show the rambling mansions that occupy this hillside, the colonnaded terraces and multiple, interconnected swimming pools. But none of the area's 648 homes can be seen on Street View; the whole area is a blank.

The same is true of another suburb with azure skies and swaying palm trees, but in an altogether different price bracket. To get there you need to scroll across that blue orb and drop down in the Indian Ocean to the capital of the tear-shaped island of Sri Lanka. If you hover high enough above Colombo, you'd be forgiven for thinking that you could jettison your cute cursor anywhere: the whole city glows blue, a sure sign that it's all on Street View. But go in closer, and you'll find that those areas with the dingy tin roofing, the unplanned webs of lanes that are squeezed into the leftover and unwanted bits of the city, are not covered. In the central district of Wanathamulla, having gone halfway up Veluwana Place with its lush trees, strolling men, bicycles and open-sided taxi vans, Street View abruptly ends. When you zoom out it's clear why: the slum beyond, like every slum in every city of the world, has not been prised open by one Google's fleet of travelling cameras.

On a global scale, the vast bulk of Africa has yet to be covered, the exceptions being South Africa and Botswana. Most of Asia has coverage limited to a few hotspots, like tourist attractions. It won't be long before all this will change. The speed with which Sri Lanka has recently been covered, with Street View now available along over 30,000 miles of its roads, shows that this technology is now worldwide. Sri Lankans seem delighted that so much of their country has been captured. The news that the great task had been completed was greeted with excited tweets and chat room comments like 'So proud'. Being on Street View is a way of showing off Sri Lanka's modernity: it's about proclaiming that the country, which has a large and active tech community, is not going to be left behind and is intent on joining the networked world.

For the moment, though, Sri Lanka is more the exception than the rule. And even in the West there are roadblocks. Apart from the odd scandal about scooping up private internet data, by and large people in the UK and US are relatively indifferent to the privacy issues that are raised by Street View. These issues are taken much more seriously in Germany. Germany has had a heated debate about the ethics of Street View, and to this day has very limited coverage. Former German Foreign Minister Guido Westerwelle went so far as to say that he would 'do all I can to prevent it'. Hundreds of thousands of people have opted to have their homes blurred out, a volume of opposition that seems to have swayed Google into thinking that it's just not worth keeping the cameras rolling.

The off-screen places of Street View are not only at the extreme ends of income, but this is where the most resilient of the hidden zones lie. Their stories tell us a lot about the changing relationship between wealth and visibility. It is rumoured that among the residents of Hidden Hills are famous names like Justin Bieber, Miley Cyrus, Jennifer Lopez and Kim Kardashian. These days the wealthy – famous and non-famous alike – don't want to be seen. Increasingly wealth and secrecy go together. It has started to seem like a natural pairing but, in fact, it's a recent phenomenon. 'There was a time when people really flaunted their wealth; now they don't,' says David Forbes, the London-based estate agent who offers a 'discreet and specialised client service'. 'People's priorities over the years have shifted,' he adds; 'Now right at the top of the list it's security.'

In *A Burglar's Guide to the City*, architectural critic Geoff Manaugh tries to pinpoint this new culture of invisibility. He calls it a 'kind of urban-scale nondisclosure agreement', allowing certain groups to be 'Out of sight – and out of reach of burglars'. Manaugh argues that 'this secrecy only adds to their property values', but warns that it also attracts the 'unwanted interest of future burglary crews'.

What does this mean for the rest of us, the visible? It seems that visibility is becoming associated with a large middle swathe – not dirt poor but not rich, either. This extraordinary technology is holding up a mirror to the ordinary. But around the corner and behind the gates, we know there are secret kingdoms. Writing in the *Financial Times*, Kate Allen argues that the 'proliferation of private public space is leading to a world with multiple levels of mapping: like a computer game, those in the know can find hidden spaces that are invisible to the unknowing mass'.

Being gated and off Street View are not the only ways you can don a cloak of invisibility. One laborious but increasingly popular technique is to dig down, creating basement levels that turn one's house into a subterranean mansion, while leaving the above-ground bits looking unexceptional. Langtry House in Hampstead Heath is a recent example. Here three bedrooms, three bathrooms and a cinema room have all been inserted below the house. I've not been to have a look, but the excited reports that all this digging has produced 'a Bond villain-style hidden lair' sound like hyperbole. Having access to three bathrooms doesn't make you Goldfinger. They also suggest that there is a public appetite for a new type of urban legend. The rumours about what lies behind the facade, or the high gates, tantalise us while confirming our lowly status as citizens of the observable world.

Not that lowly, of course. At the other end of the spectrum to great wealth is great poverty, which is just as invisible but for different reasons. Hidden Hills doesn't want our envious eyes pawing its beautiful things; it shuns the rest of the city. The slums of Wanathamulla are shunned by the rest of the city; no one wants to look. If they did, they'd see how many people live in cobbled-together shanties with only the most basic sanitation and often no electricity. On the banks of the river sprawls one such settlement; Street View stops at its many entrance points. All you see is a patchwork of walls with dirt lanes disappearing into the distance; a young man in shorts is squatting down next to a broken door which lies at his feet; a topless older man, his face unblurred, is hammering at something – you can see him through an empty window in a hot concrete house . . . all these glimpses, but the superimposed yellow line that indicates Street View is turning away, taking us back to more pleasant places to live.

The reason that Street View doesn't venture into the warren-like world of the slums is not because the streets are too narrow for its camera car. Many are broad enough and, in any case, the cameras can and often are carried as backpacks. It is because it is assumed that Google users don't want to see and the slum dwellers don't want to be seen. Maybe it would make us feel bad, or the residents ashamed. But before we become too worked up about Google's disappearing act, we need to consider a simple question: how do we know that the slums of

Wanathamulla are there? Because Google Earth shows them. On conventional maps they are just a blank space. Google Earth and Street View are technologies of visibility, and they show up what is not officially acknowledged. Activists working in slums across the world have been turning to Google Earth as a tool to cajole governments into acknowledging and assisting hidden communities. The movement for mapping slums began in India, Sri Lanka's giant neighbour. In the city of Sangli, for example, the slums were once just cartographic empty space. Google Earth changed that: they could no longer be denied or ignored. They now have recognised borders and a programme for rehabilitation.

The same thing is happening in Africa. In Nairobi a group of activist mappers who call themselves the Spatial Collective use handheld GPS devices to put the city's slums on the map. Gregory Warner, who has covered the story for National Public Radio, argues that Google Earth and the slum maps it has spawned have become a key development tool: 'A map can be entered as evidence in court to stop evictions,' he explains. 'It can be reprinted by international advocacy groups to raise awareness. It can be presented to city planners as a puzzle to be solved.'

In an ever more visible world, Hidden Hills and the slums of Wanathamulla are highlighted by their invisibility. It's an irony that connects two very different places. With that realisation comes another; that maybe the truly invisible places are hiding in the light. Where I live, where you live, is easily found on Street View, but who's interested; who looks? We're out there but we're unseen.

Trap Streets

Trap streets are deliberate mistakes put into maps in order to protect copyright. You probably won't know that the map you've copied and passed off as your own is baited with traps until a letter lands on your mat charging you with infringement. However, this dusty legal byway has other lives: it touches and awakens something about the fantasy of the map. That places we assume to be real – streets, villages – could be made up and then mixed in with the real world is disorientating and enthralling.

The enchantment of the trap street has been tapped into by recent movies. There is *Paper Towns*, which tells the story of a young woman's disappearance in the real-life map fiction that is Agloe, New York State. The Chinese film *Trap Street* is a more literal treatment – it's about a cartographer who discovers a trap street – but it too is a romance-led mystery. These geographical ghosts appear to be fertile territory for exploring the uncertainties of identity, place and love.

There have been other artistic responses to trap streets, quieter but more in keeping with their ineffable and unsettling quality. I'm thinking of an artist website called 'The Sky on Trap Street' that specialises in photographs of the sky taken in a variety of non-existent and mislabelled locations. Among the skies so far gathered are pictures of clouds and bright sunshine above Allesley Street, Wrexham; Galgenpfad, Bonn; Via Fabio Filzi, Mazara del Vallo; and Oxygen Street, Edinburgh.

The skies are real but the places they are photographed from are not. These non-places all derive from Google Maps, from a period before Google Maps started using its own Street View technology to compose its maps. Prior to 2012 Google relied on a Dutch firm called Tele Atlas, and it was Tele Atlas that squirrelled these trap streets into the out-of-the-way corners of Google's maps.

Trap streets are also called Copyright Easter Eggs. In the US they became redundant in 1997 when the courts decided that 'the existence, or non-existence, of a road is a non-copyrightable fact'. But they appear to be still in use elsewhere, although it is rarely clear how many there are. The idea that the *London A–Z* has 'about a hundred' trap streets is credited to a spokesperson for the Geographers' A–Z Map Company, interviewed on an edition of BBC2's *Map Man* programme in 2005. It sounds like a reliable source, but I'm not convinced by it. If it were true, we'd be hearing a lot more about trap streets. In fact, they are like hen's teeth; online chat rooms circle round the same few examples, many of which are now out of date. One London student, Maisie Ann Bowes, who tried to track them on the *London A–Z*, explains that she 'started by photocopying pages of the *A–Z*, and comparing the roads on it to the roads on Google Maps, checking one by one that the roads matched up by crossing them out'. It was, she says, 'a long, tedious process', but, as luck would have it, she found one on the first page she turned to. This was Whitfield Road in Blackheath, a fictitious road that bisects Blackheath Common. It is a very odd example of a trap street since it connects two main roads. Far from being an obscure alley whose existence or absence not many people will miss, Whitfield Road would be used daily by hundreds, maybe thousands of people. This suggests that it wasn't a trap street at all but a mistake.

On my *London A–Z* Whitfield Road is no longer there. As soon as such cases get reported, map-makers usually delete them. In any case, although the false street is probably a legal coup de grâce, more modest but less potentially annoying additions can achieve the same ends. In 2001 the British Government's map-makers, Ordnance Survey, received a £20 million out-of-court settlement from the Automobile Association, whose maps were charged with using copied material. The Ordnance Survey employ experts to comb over rivals' maps for telltale signs. In this case the trap that was sprung on the AA was nothing so crude as a trap street, but a series of design 'fingerprints'.

Map 'fingerprints' can take many forms. An old trick was to use a form of Morse code along the edges of a feature; regular-length segments that could be read as dots and dashes. The spacing of objects such as pylons and trees can be coded in similar fashion. The normal map user just sees a set of conventional symbols and thinks they understand what they are looking at. But what they have in front of them is an encrypted document. Slight but systematic alterations in geographical coordinates (such as shifting the last number by two) is another, and increasingly common, 'fingerprint'.

The relative width of roads can also be used to ensnare copyists. Such widths are always massively exaggerated (sensibly enough, because it's the roads that most map users want to see), but the width of minor roads relative to what you'd think were major ones can be hard to fathom. It is the kind of detail that copiers will blithely reproduce, innocently falling into another trap.

The would-be plagiarist should beware: maps can bite back. Fingerprinting was to the fore again when, in 2008, the Singapore Land Authority sued the Virtual Map Company for infringement, thus bringing the latter's popular online maps of the city state to an abrupt halt. Virtual Map claimed it had made its own maps and done all the legwork itself, but the judge quickly saw through this. The case was won because it was shown that Virtual Map had reproduced not trap streets but trap buildings; non-existent features that the Singapore Land Authority had placed on its maps to protect them. Just as telling, though rather more embarrassing for all parties, was the fact that Singapore Land Authority's maps had genuine mistakes on them and that these too were reproduced by the Virtual Map Company.

The most famous example of the insertion of a false place name on a map is the tiny settlement called Agloe, the town that's namechecked in the film *Paper Towns*. It was a fictional insert on a map produced by the General Drafting Company in the 1930s, on an empty road north from Roscoe in New York State. The toponym is an anagram of the names of the company director and his assistant: Otto G. Lindberg (OGL) and Ernest Alpers (EA). This 'trap town' worked, up to a point. Rand McNally's map of the state duly identified a place on that lonely road called Agloe, and the General Drafting Company threatened copyright breach. At this point fact and fiction begin to merge, because Rand McNally not only did not admit infringement, but they pointed to the fact that there was, indeed, an Agloe General Store on that spot. Hence there was an Agloe. Which is, of course, rather puzzling. Why was there an Agloe? It is because, when they were thinking about what to call their new store, recently built on that stretch of road, the owners looked at the General Drafting Company's map and saw that it already had a name. So Agloe came into existence. The general store shut down years ago, but if you go to Google Earth, the same spot is still tagged 'Agloe'.

Having surveyed the genesis of various trap streets and fictional map entities, I'm coming to think that many examples are as much about map-makers enjoying themselves as they are about protecting copyright. The appetite for tales of trap streets reflects a wider hunger for places off the map, for geographical phantoms and secrets. The hidden street suggests a kind of magic. In fact, it was in an extraordinary and magical book called *Mythogeography: A Guide to Walking Sideways* that I first learnt about them. The author, urban enchanter and walking artist Phil Smith mentions *Bristol A–Z*'s telltale Lye Close in a rambling but inspired disquisition that links the esoteric, unfinished nature of the universe with disturbances in everyday geography. It's the kind of prompt that pushes us to imagine trap streets as doors into another realm.

The popularity of the Harry Potter series can partly be explained because it crosscuts, or overlays, fictional and real places. You enter the other realm at platform 9¾ at Kings Cross, and Diagon Alley behind a pub. Today this is where we like our magic, close by but invisible, in the midst of the ordinary world (see 'Magical London'). In his novel *Kraken*, China Miéville muses on the trap street and asks, 'Was it that these particular occult streets had been made, then hidden?' Perhaps, Miéville continues, their names have been 'leaked as traps in an elaborate double-bluff, so that no one could go except those who knew that such traps were actually destinations?'

Miéville is hinting at the promise of the trap street. I want to visit Oxygen Street and Lye Close in the same way that one day I want to cast a spell or levitate an object. Geography and enchantment were born as twins, as children of earth magic and geomancy. They have gone their separate ways, but they know each other still.

The Uncharted
Congo

'Peat bog the size of England discovered in Congo swamp:
Scientists combat crocodiles, gorillas and elephants
to discover 80,000 square miles of uncharted territory'.
The *Daily Mail* headline from 27 May 2014 has echoes of
another era. The one-third of the planet that is above water
is supposed to be thoroughly mapped and travelled. So how
can a territory 'the size of England' go unnoticed? Closer
inspection reveals that the story tells us as much about the
persistence of the myth of 'darkest Africa' as it does about
the Congo. Google Earth shows vast swathes of forest cover,
but also the distinct lines of tracks, logging roads and small
settlements – even a few uploaded photographs. It's what
is under all that forest that is the real discovery. Dr Steve
Lewis, a physical geographer at the University of Leeds, who
led the expedition, explains that 'the satellites can't really see

through the vegetation, and can't see underneath the water of this huge wetland area'.

'Obviously we knew,' says Lewis, that the region 'is the world's second-largest wetland in the tropics. What we didn't know,' he adds, 'was that it held such reserves of peat'. I can't help feeling a little let down at this point: I expected a somewhat more dramatic discovery from this part of the world. But that, I think, is where the real interest of the story lies. The Congo is one of the planet's last remaining hotspots for the kinds of tales of discovery and adventure that were supposed to have gone out of fashion generations ago. It is *Heart of Darkness* territory in more ways than one. That story, first serialised in 1899, was about a voyage up the Congo River: it was also a journey into what Conrad calls 'a white patch' on the map, somewhere 'for a boy to dream gloriously over', but also 'a place of darkness'. One might suppose that the world of maps had moved on. But neither of the two countries that straddle this region has a mapping agency, private or public. For years the best maps of the region were large-scale ones produced by the CIA. More recently people have relied on GPS and other satellite sources. At a local level, and outside of the centre of the big cities, the land is not entirely uncharted but the printed maps that exist are basic.

It's not just the lack of maps that means the Congo is still being represented as a forbidding and unreachable destination. Although there has been low-level insurgency since the 1960s, the late 1990s saw the start of a bloody ethnic and commercially driven war in the country formerly called Zaire and, since 1997, called the Democratic Republic of Congo (DRC).

It has become what Jeffrey Gettleman in the *New York Times* called one of the 'the poorest, most hopeless nations on earth'. Three and a half million people have been killed in the DRC's various conflicts since 1998. There are, in fact, two Congos: next door to the DRC is the Republic of the Congo, or Congo-Brazzaville. The word 'Congo' comes from the Bakongo, a Bantu tribe that live across the region. Belgium was the colonial power in the west and France in the east. Both Congos remain desperately poor and, to add to the misery, sexual and ethnic violence are endemic across both countries.

The reasons for not going to the Congo are as plentiful as ever. This awful image plays, rather too neatly perhaps, into existing clichés about the region as a place of horror. It's also looking a little dated, masking the fact that both countries have fast-growing economies and are experiencing a building boom. Like many other regions in Africa, the Congo is currently seeing a road- and rail-building frenzy; new 'development corridors', many funded by mining companies, are being pushed through what were once inaccessible regions. And where there are new roads and railways, there are new maps. Helena Vilchez is a GIS expert for the mining giant Cominco Resources, which is developing a US$2 billion project in the Republic of Congo to dig out what may be the largest phosphate reserve in the world. She explains that 'There are no maps of the area or any digital 3D models, so each company that wants to explore the region needs to create its own cartography and topography.' That is exactly what Cominco is doing. It and other companies and agencies are creating detailed, high-resolution maps. All these new maps sometimes contradict each other, so a United

Nations 'Common Geographic Repository' based in Kinshasa is trying to harmonise them. The notion of the Congo as cartographic blank space is quickly becoming less of a reality than a fantasy, something that reflects our own myths and desires.

The taste for thinking about the region as an unmapped wilderness has a long history. Generations of men from the more affluent parts of the world have tested and explored themselves in this part of Africa. A decade after Conrad's tale, a book came out called *Beasts and Men* by Carl Hagenbeck, a German supplier of wild animals to zoos. Hagenbeck postulated that in these dark, deep reaches of interior Africa, long-necked sauropods might still be alive. The idea wormed its way into the Western imagination: from a *Washington Post* story in 1910 that announced that 'Brontosaurus Still Lives' down to the pages of *Boys' Life*, the monthly magazine of the Boy Scouts of America, which in 1992 ran a spread depicting the Congo Basin as 'a tangled jungle of trees, undergrowth, mud and water. There are no roads and few trails.' Commonly described as one of the least accessible places in the world, the Congo Basin and, more especially, the Likouala Swamp (which covers much of the area where Dr Lewis made his discovery), is the home, explained *Boys' Life*, of 'crocodiles, snakes, forest elephants, gorillas, monkeys, lizards, birds plus hordes of insects' and 'maybe something else'. That 'something else' could be any number of fabled B-movie beasts that hail from the Congo, from King Kong to Crocosarus, but turns out to be Hagenbeck's dinosaur. Unlike its rivals, many locals do claim that it does indeed exist. They call it the *mokele-mbembe*, or 'one that stops the flow of rivers': it's an amphibious lizard up to

35 feet long, with brownish-grey skin and a long, flexible neck. According to the tales, it lives in caves it digs in riverbanks and feeds on elephants, hippos and crocodiles. In fact, these jungles are claimed to be the home of an additional dino: there is also the *emela-ntouka*, which means 'killer of the elephants', another lizard colossus of the forest and swamp.

Where there are monsters there are monster hunters. There have been a couple of dozen expeditions to find these creatures, many from the last few decades. In 1980 and 1981, University of Chicago biologist and founder of the short-lived International Society of Cryptozoology, Roy Mackal, led a group into the Likouala and Lake Tele area. Mackal didn't find anything, apart from the tall tales of the locals. In 1992 a Japanese film crew tried again and managed to get some fuzzy aerial film of what true believers could interpret as an arching black shape in the water. It seems the less they find, the more they come. In 2012 Stephen McCullah, a young man from Missouri, raised US$27,000 (£21,000) on Kickstarter for a three-month, four-member expedition. McCullah's party fell out and soon came home. Once again the improbable *mokele-mbembe* slithered from the grasp of modern science. But that is its function – the mysterious thing 'out there', lost in what we thrillingly imagine is one of the world's 'uncharted' regions.

Maybe it would have been better if the Congo was, indeed, inaccessible. That way the litany of foreign invaders, starting with the Belgians and French in the nineteenth century, and most recently in the form of violent rebel soldiers from neighbouring countries, would have left it alone. It might have kept the mining companies at bay too. The remoteness that once protected the Congo's natural resources is fast disappearing. Dr Lewis explains that the reason the area he explored is so rich in animals – it has some of the world's highest densities of gorillas and elephants – is not a natural phenomenon but reflects the fact that it is an escape zone for 'animals that are often hunted in other places'. He is hoping that now that these forests have been designated a 'community reserve', and are managed by a Wildlife Conservation Society, their long-term survival will be ensured. One can only hope so. I'd guess that the myth, and the reality, of the 'uncharted Congo' will live on, in any case; it was never so much about a real place as an aspiration, a dream of the kind of adventure promised by the map's blank spaces.

Flat 2,
18 Royston Mains
Street, Edinburgh

This is a story about the many islands of secret money. Vast discrepancies between a place's size and its impact are created by the way that paper companies can be registered at any address, however humble. Flat 2 at 18 Royston Mains Street is one such place.

Royston Mains is an anonymous residential street of dull, brownish, low-rise blocks on the outskirts of Edinburgh. It was here that Irvine Welsh's *Trainspotting* was set, a narco-fuelled drama of nihilistic youth that was, in every sense, distant from the well-heeled tourist hotspots downtown. In 'real life' – though I'm not sure that's the right phrase – Flat 2 at number 18 was the address for 438 'shell companies'. It was here that about one eighth of the national wealth of Europe's poorest country, Moldova, was magicked away. When the scandal broke in late 2014, people in Moldova woke up to find that US$780 million (just over £600 million) had disappeared from their country's coffers. The Moldovan economy lost 12 per cent of its value. Huge street demonstrations erupted, seriously destabilising the country and leading to calls for political union with Romania. A tent town of protesters called 'Dignity and Freedom' took root outside the prime minister's offices. Soon Moldova's pro-EU government toppled and the ex-prime minister was arrested after being accused of taking kickbacks to turn a blind eye to the fraud.

Behind the front door at Royston Mains, all is quiet. Shown in by a thirty-six-year-old Lithuanian woman who runs Royston Business Consultancy, BBC Radio's *File on 4* found a 'small, homely office, which is simply the front room of the flat'. Asked about the Moldovan affair, the woman pleaded ignorance: 'we are nothing to do with this'. It soon becomes clear, though, that few if any questions are ever asked of those wishing to set up a 'business' at her address. 'Shell companies' exist only on paper. They are used to provide a discreet resting and transfer location for money, far away from the prying eyes of tax officials, or anybody else. They join the dots between the kind of fiscal impropriety exposed in the Moldovan affair and the ordinary world of tax avoidance. Those involved in the latter like to pretend they have nothing to do with the former, but both rely on clever accountancy tools, such as paper companies and an endless search for secretive financial boltholes.

The man accused of being behind the Moldovan business has a rapport with wealth. Ilan Shor is a young Moldovan millionaire who owns the country's main airport, football club and TV stations and is married to a Russian pop star. Despite being placed under house arrest for his role in the disappearance of the US$780 million, Shor was soon up and running again. Although health problems, such as high blood pressure, have meant that he has had to miss key dates in court, he shows no signs of slowing down and was recently elected mayor of the Moldovan city of Orhei.

Evidently, not everyone in Moldova puts the blame on Ilan Shor. But who do people in Britain blame? The reaction in the UK to the Moldovan scandal ranges from complete indifference to mild embarrassment. It came to light around the same time that the UK government was hosting an Anti-Corruption Summit in London. The UK enjoys scolding foreigners for shady practices. Yet far from being a diligent scourge of fiscal impropriety, it has the world's largest network of places – from yacht-strewn Caribbean islands to dour suburban roads in Edinburgh – where wealth can hunker down and switch off the lights.

The reason why there are so many such places in Scotland is that Scottish law does not require companies registered as 'partnerships' to disclose their annual accounts or the names of the people involved in them. Ownership is hidden under layers of non-specific 'general partners', who are often only traceable to other hideaways where their details are just as inaccessible, such as the British Virgin Islands. Because of its cheap, anonymous rental market, this part of Edinburgh has become an attractive place for shell companies to cluster. One example is round the corner from Royston Mains at 16/5

Pilton Rise, where over three hundred firms have been registered. One of these, called Performance Global Limited, has
been linked by a UN Security Council Report to the breaking
of UN arms sanctions. In a more upmarket part of town is 78
Montgomery Street, the home address of a whopping 3,500
limited partnerships.

The most unusual thing about these addresses is the fact
that we know them: they have been named and dragged into
the light. They are at the inept end of the wide spectrum of
fiscal loopholing. It isn't supposed to work like this: for the
professionals in this line of work, no news is good news, and all
hints of illegality are, to me, met with spell-like incantations:
'nothing we do here is illegal', and 'avoidance is not evasion'.

The scandal at Royston Mains Street is probably viewed with
quiet satisfaction in the better-managed hideaways. Most are
semi-autonomous microstates: from established British ones
such as Jersey, Guernsey, the Isle of Man, the Cayman lands,
Bermuda and the British Virgin Islands, to newcomers such
as Labuan Island in Malaysia. All of them would have done a
better job at a higher price than Royston Mains Street. Not that
any of them would ever stoop to hiding ill-gotten gains.

The 2015 global 'Financial Secrecy Index' ranks Switzerland,
followed by Hong Kong then the US as the world's top three
tax havens. However, if Britain's offshore territories were rolled
in with the UK, it would be number one. It is worth recalling
that the 'Panama Papers' leak, which engulfed Panamanian
law firm Mossack Fonseca in 2015, concerned shell companies set up not in Panama, but elsewhere, with over half of the
210,000 'companies' detailed in the papers being incorporated

in the British Virgin Islands. Tax havens make for a glamorous roll call, a world away from the drab blocks of Royston Mains. But these sunny, sandy islands and the small flats in suburban Edinburgh all exist under the protection of large states. Without Britain, the British Virgin Islands would have to find some other way of earning an honest living. These microstates have powerful supporters in their 'parent countries'.

It is estimated that, every year, about US$213 billion is shifted away from developing countries into tax havens, and that they cost the world's poorest economies several percentage points of their national income. Igor Angelini, head of Europol's Financial Intelligence Group, explains that shell companies 'play an important role in large-scale money laundering activities' and the transfer of bribe money. Given the former British prime minister David Cameron's championing of the anti-corruption cause, along with his unguarded remarks about Nigeria being 'fantastically corrupt', there is rich irony in the fact that it was Nigerian anti-corruption groups who wrote to him asking for action against Britain's tax havens. If countries like the UK, the Nigerian activists explained, 'are welcoming our corrupt to hide their ill-gotten gains', then the fight against corruption in Africa will always fail.

Tucked into the interstices of the global financial system, the clout of these tiny territories provides a model example of an inverse correlation of size with power. But they also open up the wider issue of how tax connects us to place. If you see your relationship to place as superficial, then you're not going to think twice about making sure the 'taxman' doesn't get a look-in. There are all sorts of ways of rationalising why it's better, for all concerned, that 'your money' – or indeed, somebody else's – should be 'protected'. But if you are not one of the free-floating elite and, like most of us, are in some way rooted in place and cannot or don't want to flee it, to escape its 'burden', then there is a cost.

It has certainly cost me. When you try to write books, as I do, it's a good idea to get an accountant. The first thing they'll tell you is to pretend you're a company. In the UK this means registering yourself at Companies House. Hey presto, your tax bill is cut by more than half. 'So, this is what is done,' I marvelled; and so I did it. I got a big, hardcover blue ring binder with gaps for all my directors and board members. After a few days of thinking I was terribly clever, I began to detect a problem. 'But it's not true,' I kept thinking. More than that, how can I look anyone in the eye again? This wasn't just about money; it was about my relationship to people and place. Is this – this home, this community – just somewhere I put up with until I get a better offer?

To be honest, I'm still not sure how to answer that. In the end it wasn't so much a moral or a political decision, but more a bodily reaction that decided it. I felt sick. I was sleepless. Racked by the realisation that my reward would be to be regarded as a sanctimonious fool, after one miserable week I asked my accountant to terminate the existence of Alastair Bonnett Limited. I'm still listed at Companies House; though the entry takes my corporate existence through to the end of the British tax year (Incorporated on 23 October 2012; Dissolved on 2 April 2013; Company type: Private Limited Company).

So that was the end of my little dip into tax avoidance. A gesture; a soundless nothing. Big corporations, persons of wealth, will continue to pump their money into 'vehicles' and 'shells' and all sorts of other sealed and empty things. And I'm sure they sleep deeply and well. But, at least some of the time, so do I.

Spikescape

It's a quiet afternoon in Newcastle, and having made my way across one traffic lane, I'm now somewhere I've been to often before – the central reservation, the median strip. The absence of vehicles should make the final leg easy, but no; there's that familiar jagged unevenness underfoot. I've stumbled into no man's land; a non-place that hates pedestrians. There's no elegant or easy way forward. I adopt bow legs, then tiptoe; take teeny steps amid the snarling, toothed brickwork.

All I want to do is cross the road, but now I'm a twisting thing caught on urban flypaper. I have only myself to blame: I have a weakness for shortcuts, which means that I keep finding myself in the spikescape. The slightest deviation from the prescribed path across the prescribed surface and suddenly you're in it; the landscape hisses and spits, 'Go back and go the right way!'

Spikescapes are designed to channel pedestrians and move on loiterers and skateboarders. They started popping up about thirty years ago, and are now so common that they rarely attract attention. Only the most egregious example – the small metal spikes designed to prevent the homeless from bedding down – stir any passion. But anti-homeless devices are just one product from the catalogue of deterrent paving: anti-pedestrian cobbles, ribs and nodules have been sprouting under our feet like angry mushrooms with a vendetta against the human race.

I've been trying not to get drawn too deeply into the subject of deterrent paving. I have just listed the most interesting types to my partner – tiny pyramids are my favourite – as we drive to the coast, and it turns out she isn't as interested in the topic as I initially thought. Conversationally this has left me with unspent reserves, as I've also been researching the manufacture of bench and ledge studs – 'We manufacture cone shaped studs which can be used on window sills and wall tops to stop people sitting where there [sic] not wanted' – and want quite badly to talk about Chinese benches with retractable barbs.

They started life as an art installation by the German sculptor Fabian Brunsing, who created what he thought was a protest piece about the evils of commercialisation. His bench has a coin slot that you have to keep on feeding in order to stop spikes jabbing you in the rear. Indifferent to its critical intent, Brunsing's invention was ripped off by Chinese officials, who installed coin-operated spike benches in Yantai Park in Shandong Province. 'We have to make sure the facilities are shared out evenly,' a park official explained, 'and this seems like a fair way to stop people grabbing a bench at dawn and staying there all day.'

The prospect of being impaled turned out to be an effective way of keeping bench use to a minimum. It is an extreme example, but the same logic is at work in many urban spaces where you are prodded and humiliated into obedience. It's not just literal spikes that create hostile architecture. Spikescapes are edgy, nervous, 'move along now' type places where your nerves are jangled even if your heels aren't. Other deterrents include water sprinklers designed to shift dawdlers, as well as a range of specifically anti-teen devices, such as 'mosquitoes' that emit a high-pitched squeal that only youngsters can hear and a form of fluorescent-pink lighting that highlights spotty skin. Making spaces hard to be in is not only a question of putting things in, but also of taking them away. Benches, toilets and dustbins are often hard to find in the city because they have been removed in anticipation of their misuse. The same is true for bushes and trees, chopped down and removed because they are potential hiding places for thieves and rapists.

Spikescapes are difficult, defensive, vigilant places; they are wide-ranging and far-reaching, but the star exhibit remains the anti-homeless spikes. They shot to fame in June 2014 when a snap of a small patch of inch-high metal studs installed in an alcove of an apartment building in south London was shared on Twitter. The post originated from a user called EthicalPioneer, who added the text 'Anti-homeless studs. So much for community spirit :('. Over the next few days the tweet went viral and became a worldwide discussion point. Since then, various actions against the spread of anti-homeless studs have been taken. 'A mother incensed at the installation of anti-homeless metal spikes outside a Manchester building has hit out at the owner – with cushions', runs one BBC headline, while CTV in Montreal reports a similarly hostile reception: 'It's a disgrace. This is not the kind of society I want to live in, and when I noticed that happened I want to make sure we kick that out.' In both cases, the spikes were removed.

I suspect that all this public ire about anti-homeless spikes is channelling a submerged and more complex disquiet about the increasingly regimented and disciplinarian nature of public space. It's more complex because these things are there for *our* safety: we're being looked after; someone, somehow, cares. One geographer who has tracked down this new urban realm with analytic precision is Steven Flusty. In *Building Paranoia* he has classified five different types of 'interdictory space', by which he means 'spaces designed to intercept and repel or filter would-be users'. Flusty's list has a poetic quality and runs as follows:

> Stealthy space – space that cannot be found
> Slippery space – space that cannot be reached, due to contorted, protracted or missing paths
> Crusty space – space that cannot be accessed, due to obstructions such as walls, gates, and checkpoints
> Prickly space – space that cannot be comfortably occupied
> Jittery space – space that cannot be utilized unobserved

Flusty connects the spread of deterrent paving and the rise of gated communities, linking small-scale impediments to walkers and larger-scale processes of surveillance and fortification.

Back on the anti-pedestrian paving on that ungenerous central reservation, I'm certainly getting my share of the slippery, crusty, prickly and jittery. They sound like the seven dwarfs' unlucky cousins, but they describe my predicament perfectly. The irony is that the only reason I've been snared is my determination not to use the dedicated pedestrian passage that tunnels under the road. I'm not prepared to go down there. That's not only because it will take five times as long but also because it is a gloomy, filthy place. The lighting, if it's on at all, flickers, and there's a permanent stench of urine. The most potent hostile architecture in this part of town is not the deterrent spikes – it isn't the stuff that is *meant* to deter us – it's the pathways and tunnels that were once thought to be offering an attractive and safe way home. The subway is one of many left over from the 1970s across the city; a decade when urban planners must have imagined that humans were evolving into small-brained rodents who could be channelled along mazes and culverts (see 'Skywalks'). It was the kind of perverse misjudgement that helps explain why we're now so indifferent to a dehumanising landscape. Spikescape is the unwanted and malicious child of an older generation of civic planners who eradicated the ordinary and humane pleasures of the street from our cities.

So now I'm across; there's a slight loss of dignity, but I'm used to that. I still have to pass the subway's smelly maw, and I'm reminded how much worse it could have been; better to face any number of spikes that descend into that dungeon. So I guess I'll be back on the spikescape very soon; it doesn't want me, but it will have to try harder if it wants to keep me off.

Yulin Underground Naval Base, Hainan Island

Yulin Naval Base is a secret cave base blasted out of cliffs on the southern coast of China's Hainan Island. It was first identified in 2008, and according to the US military is 'large enough to accommodate a mix of attack and ballistic missile submarines and advanced surface combatant ships'. Hazy images from satellites show two large tunnel entrances into a mountain. It sounds like a set from a Bond movie. It even has a racy set-up plot that could be worked in – the 'Hainan Island Incident'. Back in 2001 a US spy plane collided with a Chinese fighter jet near Hainan, resulting in the death of one Chinese pilot and the capture and interrogation of US crew members. Hainan Island, which is about the size of Taiwan and sits in a key strategic location right at the bottom of China, is a sensitive location, and the Chinese navy were far from happy that the existence of the new Yulin facilities was so quickly confirmed.

There is something magical about the way military power turns land into hidden territory. It's a process that touches on a deep, atavistic belief that the only real or ultimate hold we have over place is through muscle. The militarisation of place produces not just secret bases but whole regions, entire countries, blanketed with secrecy.

Naval bases buried in mountains are not that unusual. One of the world's largest is in Sweden, at Muskö. Built during the Cold War under a granite mountain, ships were both manufactured and repaired there. It also contained a hospital with a thousand beds. The underground shipyard remains in operation, though much of the rest of the base is now closed.

The Yulin cave is just one part of the recently expanded Yulin Naval Base. Looking out over the South China Sea, Yulin has been built up over recent years. However, the whole island of Hainan is dotted with military installations. The Chinese military regard it as a key jumping-off point for securing their power across the wider region, including in the defence of the disputed and distant islands Paracel and Spratly Islands (see the chapter 'The New Spratly Islands'). Its naval bases, along with its six military airfields, make it the centrepiece of the 'Second Island Chain' strategy, an arc of potential intervention and control that circles from Japan and Guam in the east round to Australia and Indonesia.

Thus, although it's true that there is a new 'secret military base' at Yulin, it is also misleading. For the whole island of Hainan functions as a secret military base. The militarisation of the island means that its landscapes are primed for military use. In this sense, the roads, towns and fields of Hainan are all hidden military facilities. In militarised societies like China, apparently civilian places have a provisional and mask-like quality. At any time the trucks could roll in, and suddenly the real purpose of the surprisingly large girth of the road, the odd hard shouldering and many other unnoticed features designed to accommodate military vehicles and personnel, become visible.

That's not just true of China. Many apparently ordinary roads in Europe or the US have some kind of military function or aspect. But it is an overlay that is especially common in China because of the size and political and social role of its armed services. The Chinese military is estimated to be around three million strong. Perhaps even more significant is the way the military interlocks with the political establishment creating a 'garrison state'. It's a phrase that was first used in 1941 by the American political theorist Harold Lasswell to describe the rise of a new kind of society founded on 'the supremacy of the specialist on violence, the soldier'. The garrison state is formally egalitarian but also authoritarian, continually enforcing and expanding a strict regime which combines and confuses patriotism and militarisation. It is, I think, a process that is equally founded on geographical claims. The Chinese military exists to defend territory, and it has extended its power by shaping the landscape to its own uses, co-opting and transforming the whole country into a mosaic of military bases or proto-bases.

Over recent years, keeping a lid on all these hidden landscapes has become something of a preoccupation in China.

The State Bureau of Surveying and Mapping, which many regard as an arm of the military, has been trying to re-establish itself as the sole cartographic authority in a country that now has a variety of internet map sites. New guidelines demand that maps can only be hosted by sites that have no history of security breaches, and cannot be disseminated outside the country. Unauthorised and illegitimate disclosures can lead to a ten-year jail term. In a curious twist on these sensitivities, the Chinese authorities accused Coca-Cola of 'illegally collecting classified information with handheld GPS equipment' in Yunnan Province. Companies such as Coca-Cola create maps to help with logistics, and they hotly dispute that they've done anything illegal. But the case showed just how much of China is considered to be militarily sensitive. Explaining the prosecution, one anonymous official in the regional mapping office rationalised that unauthorised mapping 'could jeopardise our nation'. It is a point confirmed by Li Mingde, the deputy director of the Yunnan mapping bureau: 'Mapping information can be used by enemies. So it must be restricted.' The periodic banning of Google Maps in China suggests that these 'enemies' may be as much internal as foreign.

We've come a long way from Yulin's secret submarine base. But it seems that delimiting that base, drawing a boundary around what is a secret military landscape and what is not, is no easy matter. In a militarised 'garrison state', every landscape has a hidden face. The garrison state presents us with a very particular argument about place: that it is ultimately won and lost by force. Underneath all the laws and agreements, territory isn't owned, it's occupied. To put it in terms Mao might have approved of, place comes from the 'barrel of a gun'. It's the knowledge of this brute fact that helps explain why, far from being resented, the military's territorial power is a source of pride and reassurance for most Chinese. The People's Liberation Army doesn't need to conscript. There are plenty of volunteers willing to obey Article 55 of the country's constitution that 'It is a sacred duty of every citizen of the People's Republic of China to defend his or her motherland and resist invasion.' Western journalists visiting Hainan have found that their enquiries about Yulin are met with defiant and patriotic silence. China's hidden bases are a shared secret.

Under Jerusalem

In January 1992 Dr Glock, an American biblical archaeologist, was shot dead by a masked gunman near the campus of the Institute of Archaeology in the West Bank. The Palestine Liberation Organisation wasted no time pointing to the hand of the Israeli secret services, snuffing out a scholar who had 'contributed with his technical research to the refutation of the Zionist claims over Palestine'. It was rumoured that Glock had made discoveries that undermined the official version of what lay beneath the city's streets.

The mystery of Glock's death has never been solved and the battle rages on about what lies under Jerusalem. I found this out a few years before he was shot. It was the summer of 1989 and I was a first-year PhD student attending the International Conference of Historical Geography, hosted at the Hebrew University of Jerusalem. This was my first academic conference and I was very excited. I was even deeply impressed by how pitilessly dull the lectures were and how long people could sit listening to them, on and on, without mercy. I'd arrived! This was academia. It did seem a little weird that the podium was flanked by rows of huge Israeli flags. But I remained sealed to my plastic chair, studiously fixed on one white-haired professor after another laboriously sawing their way through the knottier details of topics such as 'Male and Female Systems of Land Tenure of the Upper Middle West'. Every afternoon half the audience, which was pretty much all male, white and, to my twenty-one-year-old eyes, ancient, would change into long shorts and go out into the unforgiving sun, lining up like aged schoolboys for an organised 'field study' trip to Masada or Jaffa or wherever the coach happened to take us.

One afternoon we went to see the new discoveries under Jerusalem. We trooped down into a mercifully cool subterranean chamber and strip lights flickered on to reveal our silent, respectful faces and some low yellow walls, the homes and storehouses of Jewish families who lived in this city many thousands of years ago.

'Rumours are circulating of our Palestinian contact's anti-propaganda tour – he was supposed to go public in the last session today but never did.' This is an entry from the diaries I kept back then, though the diary from that summer is like a pile of scraps. It was written on any bit of paper I could find: receipts, the backs of bus timetables, a lined exercise book half-filled with a schoolchild's Hebrew which I found in a thorn bush. Every year I think it is time to throw these diaries out, but can't quite do it. They were, after all, a lot of work. I can barely make out my hurried scrawl, but some passages are clear: on this sheet, for example, I report that the conference was 'invaded by Palestinian geographers who wanted to take us on a field trip beneath the city'.

My diary isn't reliable. Much of it was written when I'd had a bit too much to drink, which I tried to do most evenings. But that account is bringing back a memory. I can recall an argument at the back of the conference hall: a man standing up and protesting, some pushing, a moment of tension and a moment of choice. We were being asked to go off, right now, on a very different field trip to see the Palestinian story of what lies beneath the city. But inertia and diffidence won the day, a deference to authority masked as politeness. The delegates remained in the hall; there were coughs and whispers. The chair suggested we carry on with proceedings, so we did.

Who, in an old city that has seen change after change, gets to say *stop digging*, that's the layer that matters, those are the foundations that count, not the ones below or the ones we've just dug through? Dig below today's Jerusalem and you find an Islamic one; dig below that and you find a Jewish one, or, depending on where you look, a Christian one, or a pagan one. And that's to presume that religion is the best way we should tag these strata, which seems unlikely. This city has seen a long succession of rulers – Jebusites, Babylonians, Greeks, Persians, Romans, Byzantines, Mamluks, Ottomans and the British – the story of any of whom is easily lost by the way the past and the present in this part of the world are chunked up into religions. These dilemmas are not unique to Jerusalem. Most communities, certainly most nations, turn to archaeology as a way of cementing their legitimacy; of saying we have a right to this place because we have been here for a long time. The historian

Eric Hobsbawm's favourite example was a book called *Five Thousand Years of Pakistan*, about a country only seventy years old. It's odd how often that number, five thousand years, crops up in national histories in the Middle East and Asia. Chinese, Indian, Egyptian and, indeed, Israeli historians have all plumped for five thousand years. Some of the really eager ones go for ten thousand. I guess these are the grandiose spans that root a people in a place; make them seem as natural and permanent as the mountains. Thousands of years are dangled before us as a way of ensuring our submission before claims to indigeneity. They are supposed to give us a bit of a tingle, a respectful awe that we are in the presence of true natives. Yet there's a touch of paranoia about these numbers: they are too much, a little cracked, like a child whirling round after one too many taunts and shouting something absurd.

In many towns and cities what lies below is not just as important as what sits on top it is *more* significant. The surface layer of a city is ephemeral: roads and malls go up or come down in the course of a few weeks. Below ground has a pristine permanency, an inviolable solidity. It promises a weight of evidence which is especially important to societies that are racked by self-doubt. It is doubt that drives the digging. In the case of Jerusalem, that includes doubt about religion as well as doubt about the nation. Archaeology in this city took off in the late nineteenth century as Europeans searched for evidence of the literal truth of the Bible. Why did they feel the need to search, to find proof? Because they lived in an age of increasing doubt, of increasing uncertainty about religion. The leading organisation of this movement was The Palestine Exploration Fund, which is still resident at 2 Hinde Mews, Marylebone Lane, London. It was founded in 1865 and recruited such luminaries as Charles Warren, former London Commissioner of Police during the time of Jack the Ripper. He was employed to go and prove the veracity of the Bible's stories through scientific study. And he did surprisingly well. Warren excavated the Temple Mount and a tunnel which is now known as Warren's Shaft. The Old Testament tells us that David, King of Israel, took Jerusalem from the Jebusites by getting his nephew, Joab, to crawl along a water shaft and launch a surprise attack on the city. Warren's discovery was taken to be Joab's tunnel and, hence, proof of the story.

Sieving the earth for proof became an even more determined exercise a hundred years later, after Israel's capture of Jerusalem's Old City in 1967. Archaeological excavations got under way almost immediately, using a ready and eager supply of soldiers and students. By this time the focus had switched towards finding tangible evidence of Jewish occupation and the struggles of earlier Jewish states to survive and flourish. These ambitions come together at the Burnt House. Discovered under a layer of ashes about six metres below street level, archaeologists concluded that 'the house was destroyed during the destruction of Jerusalem at the hands of the Romans, in 70 CE'. It was identified as the home of the Kathros family, a clan who have a somewhat unflattering write-up in a key compendium of Jewish doctrine, the Talmud: 'they were all high-priests, their sons were the treasurers, their sons-in-law were the chamberlains, and their servants would beat us with rods'.

This is the kind of human angle that we increasingly want from archaeology. We want a living, breathing communion with the past. But who is being brought to life and who is remaining dead? All across Jerusalem the bones of ancestors are reanimated into competing claims and warring armies. No site is more contentious than the Temple Mount. Palestinians used not to worry that Jews regarded the Temple Mount, known to Muslims as the Haram esh-Sharif, now home to the Dome of the Rock and Al-Aqsa Mosque, as the site of the First Temple. It was readily admitted this was a possibility. But in 2000 Palestinian leader Yasir Arafat reflected a new and hardening mood: 'I will not allow it to be written of me that I have . . . confirmed the existence of the so-called Temple beneath the Mount.' An archaeological zero-sum game began to be played: it's either yours or ours; it cannot be both. Building spoil trucked away from the Temple Mount by its Muslim curators has been suspected by Jewish activists as evidence of historical sabotage. Thus began the Temple Mount Sifting Project. This back-breaking and ongoing task, sifting through dumped rubble and soil, achieved something of a eureka moment when it turned up a tiny clay seal bearing Hebrew letters interpreted as reading 'lyahu son of Immer'. The Immer were a priestly family, another of whom, Pashhur, is mentioned in Jeremiah as 'Chief officer in the house of God'. Once they emerge into the light the fragments that once lay quietly under the city become weapons in the region's ceaseless religious animosities. In 2010, the Palestinian Authority ramped up the tension again by declaring that the Western or 'Wailing' Wall, which abuts the Temple Mount, wasn't a Jewish holy site but part of the Al-Aqsa Mosque: 'This wall was never part of the so-called Temple Mount, but Muslim tolerance allowed the Jews to stand in front of it and weep over its destruction'.

It's odd how what is beneath our feet can weigh us down. That's how I remember Jerusalem. If, that is, I remember it at all. I get it mixed up with other places and, despite my initial efforts, I wasn't concentrating. My diary makes it clear that my main concerns were finding a drink and avoiding the other delegates. When the conference was over, I stayed in a youth hostel on a hot hillside outside the city. 'Lots of German girls with large wooden crucifixes on their chests'; 'Walk miles with bag on sunburnt shoulder full of water and Vermouth'; 'Got the 193 bus to Jericho today but miss the stop because it was too small.' I guess I must have relaxed a little, letting the anger and anguish of Jerusalem wash away.

Doggerland

I've come to the village of Covehithe in Suffolk because it is the fastest-disappearing place in Britain. The low, sandy cliffs are falling away into the sea, sometimes by several metres a year, and they have been doing so for years. What was a Roman settlement, and then a medieval town, is now just a few houses and the walls and tower of a large ruined church. There's also an overgrown tarmac road heading to the shore; a road that breaks up and stops abruptly in a dense patch of briar before tipping over a vertiginous drop.

Everything's still. It is a windless day. The sun shines weakly on a quiet sea. This edge, this village and coast will one day join the hidden territory under the water. At the end of the road, looking out to sea, the name of this expanding domain came back to me – Doggerland – along with the strange realisation that, as I gazed over that blank sea-bound kingdom, I was feeling something like nostalgia.

The label 'Doggerland' was invented in the 1990s, and stems from the doughty little Dutch fishing boats, or doggers, that once plied the North Sea. It's a modern name, but it's a compelling one. On the crumbling, uncertain fringe of England, Doggerland, its rivers and forests, its history and denizens, draws us down.

The 'land' part of the word is its oddest component. There is no land there now. It is a distorted echo from the deep past, from a time when there was no North Sea or English Channel; when Europe was a much bigger place. Doggerland was inundated by the rising sea as glaciers melted and retreated, at first incrementally but then catastrophically in a great tsunami. This was a world overtaken by disaster; people, villages, vanished, stopped in time. The poignancy of this undersea Pompeii is compounded by the knowledge that many parts of the world, including much of Suffolk, are one day going to join it.

Let us carry on walking along that dead-end road in Covehithe. As we do so, let's step back in time ten thousand years. What do we see before us? It is a lush, low-lying landscape of braided rivers and a seemingly endless forest; of oak, elm, birch, willow, alder, hazel and pine. Through the trees we catch sight of a wild horse and a herd of bull-like horned animals, aurochs. At intervals across this dense green plain, thin trails of smoke spine the breathless air. They are coming from cleared areas as well as from riverside settlements. Moving closer, we see a few thatched huts and a central firepit. This is a small community, an extended family of maybe two dozen people. They are working at something; sharpening and carving notched barbs into deer antlers. In many thousands of years time, in 1931, 25 miles off the coast of Norfolk, one of these sleek harpoons will be pulled up in the nets of a fishing trawler; the first tangible clue of a lost civilisation.

The villagers are talking and singing but not in any language that's easy to identify. Its closest modern-day relative is Basque, the oldest living language in Europe. What is clearer is that, although they are largely nomadic, they know this place intimately. These are not newcomers: their ancestors have been living here for thousands of years. For some of that time they would have shared this territory with those who were here well before them. A chunk of Neanderthal skull pulled from the North Sea in 2009 has been dated to about 58,000 BC.

To the north we can make out a speckle of small lakes and one huge one, a white mirror on the horizon. It's shining so brightly today that it's hard to see its furthest shore. It's a freshwater lake, but at 75 to 105 miles wide it's big enough to look like an inland sea. Today this low-lying underwater basin is called the Outer Silver Pit. Away to my south are some deep valleys and a single great rock, a site of ancient ceremony and pilgrimage; an undersea feature that contemporary navigators call the Cross Sands Anomaly. Straining my eyes, I think I can make out the faint outline of higher ground far to the north, also thickly forested. Perhaps this is the range that was once thought to be the heart of Doggerland, the Dogger Hills. But I blink and it's gone; it's too far for me to be certain. Recent studies lean towards the idea that the Dogger Bank, a vast shallows at the heart of the North Sea that lies about 45 to 60 feet beneath the surface of the water, was laid down after this plain was inundated.

This is the heart of Europe; a fertile landscape rich with opportunities for hunters and gatherers. It is a peaceful scene, settled and rooted; but it will not last. The higher places will become the last redoubts of the Doggerlanders, and then they too will be drowned. Over recent years archaeologists have been mapping Doggerland, producing surveys of its rivers and shores. They have also been trying to imagine how people responded to this traumatic event. It seems that these Stone Age people were taken by surprise. In *After the Ice*, Steven Mithen says that 'some day close to 7,000 BC' a catastrophic event suddenly destroyed their world. A massive landslide over 600 miles to the north triggered a tsunami: 'white, stony sand buried everything as far as could be seen to the north and south'. The impact was overwhelming. Mithen describes the moment disaster struck: 'Many kilometres of coastline are likely to have been destroyed within a few hours, perhaps minutes and many lives were lost; those hauling up nets from canoes, those collecting seaweed and limpets, children playing on the beach, the babies sleeping within bark-wood cots.'

Another archaeologist who has worked on Doggerland, Clive Waddington, argues that the consequences of the flight of the Doggerlanders would have been felt across the wider region, with communities defending themselves against the refugees. Other responses to the rising tides were waterside shrines or offerings. Flints and antler hoards recently found in the construction of Europoort harbour in Rotterdam have been interpreted as having a religious function; these treasures being carefully placed on the shore to placate the angry spirits of the sea, a kind of climate-change shamanism.

I walk back up the overgrown road in Covehithe and take a long, looping path down to the empty beach. Delicate waves drag at the shingle. What is there here? Nothing? There is no memory of Doggerland; it feels more distant than the furthest star. But as we learn more about it, and as it is mapped and imaginatively possessed, something rather odd has happened. Its story has become ours, and we have begun to care about these long-forgotten ancestors. Down here on the quiet shore I have a gnawing sense not just of a lost past but of loss itself, of grief. We face the same basic challenge: of hanging onto our land as the climate shifts and the water rises. We are joining them, these lost people; sitting down beside them, where we can share stories of survival.

The New Arctic

The top of the world is disclosing secrets that have lain hidden for hundreds of thousands of years. As the ice disappears, new rivers are flowing; islands are breaking the waves; shipping routes are opening and geopolitical disputes heating up.

This is all happening at the same time as imaging techniques are allowing us to peer beneath the ice and map out the Arctic's buried landscapes. The most spectacular example is from the Arctic's largest island; a 460-mile mega-canyon, longer than the Grand Canyon and of a similar depth, has been found cleaving the heart of Greenland. We only learnt about its existence in 2013 when NASA's airborne radar picked it out deep below the ice. The 'Greenland Grand Canyon' lies under some of Greenland's thickest ice, which can be almost two miles deep.

'It is quite remarkable that a channel the size of the Grand Canyon is discovered in the 21st century below the Greenland ice sheet,' says Michael Studinger, one of the NASA scientists, adding, 'It shows how little we still know about the bedrock below large continental ice sheets.' The truth of this was again borne out when another canyon was discovered, although this one was at the other end of the world. It cuts down through the bedrock of a little-explored region of Antarctica called Princess Elizabeth Land. It is even bigger than its Arctic cousin; at 620 miles long it dwarfs any rival. It is still only tentatively mapped, but it is thought that it may be connected to a huge, 483-square-mile subglacial lake.

These mega-canyons were once the heart of river systems, and they still provide a channel for subglacial meltwater to reach the ocean. Scientists are interested in how much meltwater reaches the open sea because it affects global sea level rise. Another NASA discovery was made in 2017, again in Greenland and again changing the way we look at the Arctic. Channels in the ice were found running down to the sea along which gushed meltwater fed from huge caves. The caves, called aquifers, were themselves only discovered in 2011. They are formed in the firn layer. 'Firn' is a Swiss German word for 'last year's', and refers to a band of snow that has been squeezed into a dense mass. We now know that firn aquifers cover around 8,500 square miles of Greenland and that the water within them never freezes, as the snow above acts as a blanket keeping the temperature up.

The discovery of these Arctic fluvial systems comes at a time when this once-unvisited landscape is being wrenched open by human activity. Much of the heavy lifting has been undertaken by Russia's Northern Fleet. In 2015 they declared that they had discovered nine new islands as well as 'five straits, seven capes and four bays'. They reported that the scale of the discoveries was such that it required the 'reprinting of existing maps, guides and manuals for navigation'. The largest of Russia's new-found islands was discovered in the Gulf of Borzov, and is one mile long and 1,968 feet wide.

These finds are emerging, in some part, because of 'glacial bounceback' (see 'Bothnia's Rising Islands') but largely because of the retreat of the ice. The Northern Fleet's staccato accounts of its findings rap out the facts and invariably tie their emergence to vanishing ice. For example, 'Glacier on the eastern shore of the Gulf Vilkitsky departed to the east by 2–6 km, forming a cape and two islands. The lip of Glazov glacier melted and shifted east by 2–3 km, in consequence, in the opened bay formed two capes and two islands.'

New islands have also been recorded emerging from the Svalbard archipelago; the southern part of the main island of Spitsbergen is dividing off into a separate island. The western side of the Arctic has also seen new islands emerge, although much of the effort here has been expended on securing an ice-free route over the top of Canada. For centuries mariners have looked for the Northwest Passage. It took Roald Amundsen three years, from 1903 to 1906, to become the first person to navigate it successfully. At times the water Amundsen sailed was only three feet deep, so back then the route was of little practical value. Today the Northwest Passage can be sailed all year round. In 2016 the cruise ship *Crystal Serenity* took a thousand tourists on a leisurely reconnoitre of Amundsen's route. However, the channels above Canada are complex;

many are still shallow, and while the US and Europe claim that the Northwest Passage is an international strait, Canada says it is not. Hence other options are being actively planned. A fantastic new possibility is seizing the imagination, the Transpolar Sea Route: a shipping lane that will allow vessels to sail straight over the top of the world, connecting the Pacific with the Atlantic.

The first surface ship to sail all the way to the Geographic North Pole was the Soviet nuclear icebreaker *Arktika*. It arrived at the North Pole on 17 August 1977. According to the Northern Fleet's figures, which cover the period up to 2008, there have been a total of 77 subsequent voyages: 65 Russian, 5 Swedish, 3 America, 2 German, 1 Canadian and 1 Norwegian. The Russian commentary adds that only nineteen of these trips were done for scientific reasons; 'the remaining 58 were for the entertainment of tourists'. All but one of the trips were undertaken in the summer. The exception was the winter journey of the Soviet nuclear icebreaker *Sibir*, which thrust its way through near-maximum thicknesses of Arctic ice.

The Arctic Ocean is being covered with lines of traffic. It is being normalised; like other oceans it is being plotted with trade routes. The first transect of the Arctic Ocean was under-taken in 2004 as a collaborative effort between two icebreakers, the Canadian *Louis S. St-Laurent* and the US *Polar Sea*. No ordinary ship could make the journey, but when that finally happens the Transpolar Sea Route will revolutionise world transport. Travel times for goods taken between East Asia and the Atlantic will plummet. It will also render the Suez and Panama Canals semi-redundant.

There is still a lot of uncertainty about when the Transpolar Sea Route will become viable. Ice information and weather forecasting in these high waters are still unreliable, and estimates about when the route will be sufficiently free of ice vary. Some say it will be as early as 2030; other estimates push this back to 2050 or later. Shipping companies are already incorporating the route into their long-term planning but there are other Arctic passageways, particularly along the top of Russia – the Northern Sea Route – that may prove safer and more reliable.

The Arctic has more than one type of magnetic attraction. It has been estimated that about 22 per cent of the world's undiscovered and recoverable resources are in the Arctic Circle. These untapped oil and gas reserves mean that every Arctic power wants its own slice – the bigger the better. As a consequence the region's undersea geology has been politicised. One of the culprits here is the innocent-sounding United Nations Convention on the Law of the Sea. It allows a country to claim a patch of ocean beyond the normal 200 nautical miles if it can show that it is part of its continental shelf. The UN convention defines continental shelves as the 'natural prolongation' of a nation. Geologically this is a very odd idea and, politically, it is a recipe for conflict. Continental shelves usually extend from multiple nations and their boundaries are hard to define.

Denmark has recently presented a claim that much of the North Pole is on the same continental shelf as Greenland, and thus theirs. Christian Marcussen of the Geological Survey of Denmark announced that 'The Lomonosov Ridge is the natural extension of the Greenland shelf.' The Russians claim the same undersea feature, saying it is a natural extension of their own shelf. At its highest point the Lomonosov Ridge lies at a depth of 3,129 feet. It is a colossal 1,120-mile-long upland that runs straight across the North Pole. If it were on land it would look like a mountain chain, with peaks rising up to 12,140 feet into the sky. The Lomonosov Ridge was discovered by Soviet explorers in 1948, and neither the Danes nor the Russians appear willing to share it. Drilling their point home, in 2007 a Russian mini-submarine planted a titanium flag 13,780 feet beneath the North Pole. In 2015 Russia formally claimed almost 400,000 square miles of Arctic territory.

The Arctic is giving up its secrets; new landscapes and new opportunities are breaking the waves. If the squabbling continues, and efforts to establish the Arctic as a common inheritance and as an international conservation area cannot bear fruit, it will soon be witness to a twenty-first-century gold rush. The release of huge reserves of oil and gas will further exacerbate global warming and the retreat of the ice. In future years, we may be known as the generation that gave away the Arctic, even though it was not ours to give.

Conshelf
Undersea Station

Since two-thirds of the planet is water, it is no surprise that we dream of making a home for ourselves under the sea. The most radical, plentiful and determined efforts to plan and build undersea villages were clustered within just two decades, the 1960s and 1970s. Worries about the growing human population along with funding from government and industry, created a marvellous shoal of eccentric submarine homesteads. Some of the most ambitious were designed by the French architect Jacques Rougerie, whose plans for deep-sea dwellings resemble aquatic creatures: finned and globular, they look as if they are about to flick their tail and scoot away.

It is the ones that got built – the prototypes for the coming race of oceanauts – that are the most exciting. They were small, inhabited by just a few divers, but they worked; they proved that it is possible to live underwater for extended periods.

In 1977 the indefatigable Rougerie got to dive to the deep for real in one of his own creations, an elegant, single-stalked marine mushroom called Galathée. For a while, superpower rivalry turned undersea living into a new space race. American underwater habitats and laboratories – Sealab I and II, Tektite I, II and III, Edalhab and Hydrolab – were rivalled by the USSR's 'Ichthyander Project', the 'underwater passenger complexes' Sadko I, II and II and Bulgaria's 'subsea house' Gebros-67.

Today it is forgotten that it was not just reaching the moon and other planets that caught the imagination of that generation; the oceans also beckoned. One 1968 Soviet popular science book hailed the arrival of *Homo aquaticus*.

All these schemes were inspired by the extraordinary achievements of another tireless Frenchman, the doyen of the deep Jacques Cousteau. Cousteau's Conshelf Underwater Stations I, II and III were talismans of what, for a decade or two, seemed like a bold aquatic dawn. After the success of the first Conshelf, submerged off Marseilles in 1962, Cousteau was ready for something much more ambitious. In his Oscar-winning 1964 documentary on Conshelf II, *World Without Sun*, he created an enduring record of the life of an oceanaut.

The film opens with an otherworldly sight; a 'diver saucer' flying through the deep, plumes of sand and water shooting from its propulsion jets. It glides beneath an underwater hangar and is winched up inside. We then see the saucer's crew dive to their main base, the four-podded Starfish House which had five bedrooms, a kitchen and a laboratory. Cousteau's unique and stately drawl informs us that this 'underwater village' is 'conceived not only to explore but to inhabit the sea'. We see the oceanauts, who will live on Conshelf II for a month, at leisure. In scenes that inspired Wes Anderson's comic homage to Cousteau, *The Life Aquatic*, they saunter about in jaunty swimming briefs, smoke their pipes, adjust the volume of classical music and do a spot of light hovering. At one point a smiling barber swims aboard and sets to work; in another scene, an oceanaut arrives with his pet parrot.

It is hard to imagine a similar endeavour today being so charmingly carefree. And it is easy to forget that Conshelf was a serious scientific project. Cousteau details the challenges of living in a sealed undersea unit for such a long time. Despite having electricity, air-conditioning, fresh water and television, Conshelf's six residents found that, without the normal rhythms of light and dark, their sense of time became hazy. Living with more than twice the air pressure of people on land made movement wearisome and the high oxygen content took a while to adjust to. Insouciant as ever, Cousteau remarks that the high oxygen levels 'makes tobacco burn twice as fast'. His attention to the art of smoking becomes apparent again when one of the men is preparing to depart for a week to the 'deep experimental cell', a seven-foot cabin some eighty feet below. Cousteau solemnly informs us that 'down there he will have to give up his pipe'. The deep cell has other effects: the voices of the two men stationed there turn into squeaks and 'cuts and abrasions heal overnight; beards almost stop growing'.

Cousteau's work had previously been, in part, supported by grants from French oil companies, and he was unabashed that one of the goals of his prototype communities was 'to systemically exploit the resources of the ocean'. However, siting Conshelf II down in the Red Sea off Sudan, one of the warmest and richest environments for marine life, shows that Cousteau's real passion was for conservation. He was beginning to reject his earlier alliances and forge his own path. In any case, Conshelf was also proving how expensive it is to live underwater. Governments and companies who wanted to get access to subsea resources had already realised that robotic and drill techniques were more effective and cheaper.

We don't need undersea villages to help us drag stuff off the seabed, but Conshelf always had bigger ambitions. Cousteau wanted to open up a new horizon for human habitation. And he did. The divers returned after their month-long stay in good shape, and the Starfish House and deep cabin were lifted up and sent back to France. With Conshelf II, Cousteau had also invented a new form of scientific funding: one that exploited not the seas but public fascination. As well as *World Without Sun*, throughout the 1960s he made numerous television programmes that changed the way people understood the oceans. Before Cousteau, they were dreary immensities. His programmes transformed the seas into places of wonder that deserved attention and care. The cameras were rolling again when, in 1965, Conshelf III was lowered to a depth of nearly 330 feet off the waters of Nice. That was the last of Cousteau's underwater living environments. A decade or so later, the heady days of *Homo aquaticus* began to fade from view.

Cousteau died in Paris in 1997 at the age of eighty-seven. There have been more recent forays into the deep, but not many. At the moment the world has only one undersea habitat: Aquarius, in the Florida Keys. In 2014 Fabien Cousteau, grandson of Jacques, spent thirty-one days in Aquarius, beating Conshelf II's record by one day. There was a touch of nostalgia to Fabien's visit, for it paid tribute not only to his grandfather but to a whole generation for whom the oceans were an exciting new frontier ready for colonisation.

Conshelf II has become a favourite spot for scuba divers. Although the Starfish House is no longer there, the hangar for the dive saucer remains intact and according to *Dive the Red Sea*, is 'still airtight and housing a large bubble formed by visitors' exhalations'. Other remnants include shark cages and a tool shed. When in operation, the outside surfaces were cleaned every day to stop any build-up of seaweed. Today the whole scene resembles a coral garden populated by reef fish and a multitude of grey reef, silky and hammerhead sharks.

Conshelf was not a false start, a bridge to nowhere, but a working prototype. We are still at the start of this adventure. Living under the sea is possible, that much is clear. The bigger question is, can we turn what is, after all, a hostile and deadly environment for *Homo sapiens* into an attractive option? It may be that the answer to that question is to do with scale. Though the oceans are vast, submarine life has, so far, been a cramped and uncomfortable experience. New schemes scale up undersea habitats. Jacques Rougerie is still dreaming big, and is planning a huge, roving craft, which towers both above and below the waves. The prototype is planned for launch in 2020. Even bigger is Japanese company Shimizu Corporation's Ocean Spiral. Designed to have a permanent population of four thousand, it will be self-sufficient and self-sustaining, generating its own power and fresh water. The Shimizu Corporation explains that their 'large-scale concept seeks to take advantage of the limitless possibilities of the deep sea by linking together vertically the air, sea surface, deep sea, and sea floor'. Their enthusiasm appears boundless: 'Now is the time for us to create a new interface with the deep sea, the earth's final frontier.'

It has been 'the time' before and it will be again. The desire to colonise 'earth's final frontier' will not be extinguished; not by practical difficulties, nor by the unnerving prospect of entering a 'world without sun'. The pull of the oceans goes beyond the rational. It is not necessary to go there, but we know, instinctively, why people choose to. In the warm waters of the Red Sea lies one of the coral-crusted monuments to this determination; where not so long ago some of the first oceanauts made their home, fed their parrot and drew contentedly on their pipes.

Epilogue

Sometimes I'm asked to list which of 'my amazing places' would make great holiday destinations. It's an understandable question, but I don't warm to it. Holidays should be fun and carefree, and I doubt that many of the thirty-nine places in this book qualify. My inner rebel is also yelling 'the point is to find your own'. The thirty-nine places are prompts; they are designed to make us wonder and wander.

Far from advising people that they have to troop off to one of the world's umpteen grindingly soulless airports and get a long-haul flight to find somewhere truly memorable, my advice is not to fly or drive anywhere at all. Set off on foot from your own front door and head in a new direction. Don't walk quickly or have your head down, and don't give up after half an hour. Let it happen and give it time. I'm increasingly convinced that walking is the only real form of travel: everything else is just speeding past. It can be hard work: the experience you get will not be tailor-made, not packaged, not a cliché. There will be no forms to fill in; you won't be asked to remove your belt and shoes; there will be no guidebook. Many of the places in this book started with that first step, but I could have filled every page with them.

I have corralled, from my own travels and researches, a nervy herd of unruly places. They are splendidly different, but there are themes that lash them together; themes that suggest the old view of geography as a collation of known and clear borders and established, accepted facts is disintegrating. The world exhibited here is fragmented and fragmenting; it is surging with utopian and secessionist ambitions and it harbours legions of ghosts and endless secrets. We have seen places splintering, becoming strange and felt the power of these mutations; how they touch us. Some of these places may appear remote and some commonplace, but all their stories are our stories. Geography is getting harder to read; the map is breaking up. It is an extraordinary, even magical sight, but it's also bewildering and often frightening. I used to think that this kind of glittering spectacle could be described as 'enchanting' and that what I was doing, as I captured and catalogued escapees from the map, was offering a 're-enchanted' geography. But now I'm not so sure. The forces at work have become too unpredictable to be easily or neatly summarised. We're all strapped in on a hurtling and disorientating geographical ride. We don't know where we are headed, but there is no getting off. All we can do is open our eyes and hang on tight.

Bibliography

Akiba, Shun. *Teito Tokyo Kakusareta Chikamono Himitsu* [Imperial City Tokyo: Secret of a Hidden Underground Network], Yosensha Publishing, Tokyo, 2002

Barbrook, Richard. *Imaginary Futures: From Thinking Machines to the Global Village*, Pluto Press, London, 2007

Buckles, Guy. *Dive the Red Sea*, New Holland Publishers, London, 2007

Burns, Wilfred. *Newcastle: A Study in Re-planning at Newcastle Upon Tyne*, Leonard Hill, London, 1967

Ehmann, Sven, et al. *The New Nomads: Temporary Spaces and a Life on the Move*, Gestalten, Berlin, 2015

Flusty, Steven. *Building Paranoia: The Proliferation of Interdictory Space and the Erosion of Spatial Justice*, Los Angles Forum for Architecture and Urban Design, West Hollywood, 1994

Frampton, Adam, Solomon, Jonathan and Wong, Clara. *Cities Without Ground: A Hong Kong Guidebook*, Oro Editions, San Francisco, 2012

285

Graham, Stephen. *Vertical: The City from Satellites to Bunkers*, Verso, London, 2016

Hagenbeck, Carl. *Beasts and Men: Being Carl Hagenbeck's Experiences for Half a Century Among Wild Animals*, Longmans, Green & Co., London, 1909 (reprinted 2016)

Manaugh, Geoff. *A Burglar's Guide to the City*, Farrar, Straus and Giroux, New York, 2016

Miéville, China. *Kraken*, Macmillan, London, 2010

Mithen, Steven. *After the Ice: A Global Human History, 20,000–5000 BC*, Weidenfeld and Nicolson, London, 2003

Papadimitriou, Nick. *Scarp*, Sceptre, London, 2012

Raspail, Jean. *The Camp of the Saints*, Noontide Press, Costa Mesa, 1986

Rees, Gareth. *Marshland: Dreams and Nightmares on the Edge of London*, Influx Press, London, 2013

Rensten, John. *The Edible City: A Year of Wild Food*, Boxtree, London, 2016

Rogers, John. *This Other London: Adventures in the Overlooked City*, HarperCollins, London, 2013

Smith, Phil. *Mythogeography: A Guide to Walking Sideways*, Triarchy Press, Axminster, 2010

Watkins, Alfred. *The Old Straight Track: Its Mounds, Beacons, Moats, Sites, and Mark Stones*, Methuen, London, 1925

Wertheim, Margaret. *The Pearly Gates of Cyberspace: A History of Space from Dante to the Internet*, W. W. Norton, New York, 2000

Whiting, Charles. *The End of the War, Europe: April 15–May 23, 1945*, Stein and Day, New York, 1973

Acknowledgements

I've sought and received a lot of help to complete *Beyond the Map* from many people in different parts of the world. Particular thanks to Lucy Warburton and Ru Merritt at Aurum, Mary Laur at University of Chicago Press, Jenny Page, James Macdonald Lockhart, Rachel Holland, Anna Macdonald (for research assistance) and my many colleagues and friends at Newcastle University.

Index

Page references in *italics* indicate illustrations.

289